Contents

DARWIN

WOLF

PINTA

ROCA REDONDA

PTA. ALBEMARLE

CAPE BERKELEY

CAPE MARSHALL

PTA. VICENTE ROCA

BANKS BAY

SANTIAGO

ALBANY

BARTOLOM

JAMES BAY

TAGUS COVE
PTA. ESPINOSA

PTA. GARCIA

COWLEY

BAINB ROCK

CAPE DOUGLAS

URVINA BAY

RABIDA
(JERVIS)

BEAGLE ISLANDS

CAPE HAMMOND

PTA. MANGLE

GUY FAWKES

FERNANDINA

EDEN

CARTAGO BAY

PINZON

ELIZABETH BAY

NAMELESS

PTA. MORENO

ISABELA

LOS HERMANOS

PTA. VEINTIMILLA

CALETA IGUANA

VILLAMIL BAY

TORTUGA

CAPE ROSA

SCALE (KILOMETERS)

0   1   2   3   4   5

FLOREANA

BLACK BEACH

ARCHENA

PTA. ESPEJO

GENOVESA
(TOWER)

NORTH SEYMOUR

BALTRA

· GORDON ROCKS

PLAZA ISLETS

SANTA CRUZ

· CAAMAÑO

RTO AYORA &
DEMY BAY

SANTA FÉ

KICKER ROCK

PTA.
PITT

PUERTO
BAQUERIZO
MORENA

ROCA ESTA

SAN CRISTOBAL

VIL'S CROWN

TA. CORMORANT

ENDERBY & CHAMPION

CALDWELL

GARDNER

WATSON

ESPAÑOLA
(HOOD)

GARDNER

PTA. SUÁREZ

PTA. CEVALLOS

The Galápagos Islands

Preface

This book is designed to serve as a compact field guide to the commonly encountered echinoderms of the Galápagos Islands. It is intended primarily for interested visitors who wish to explore the fascinating variety of Galápagos life in tide pools and on beaches while walking the shoreline, or those that may don mask and flippers—or scuba gear—to view the vastly different underwater world beyond the beaches. I trust it also will serve as a convenient reference book for scientists engaged in marine research.

The need to identify Galápagos marine invertebrates, and the absence of a handy means for doing so, became obvious to me when, in 1989, I began conducting annual spring censuses of the Galápagos intertidal fauna with the help of students from Washington and Lee University. To construct a provisional field guide for student use I first had to locate and consult the scattered reports and monographs of numerous investigators of past expeditions to the Galápagos. This effort quickly revealed that the published literature treated certain groups comprehensively— molluscs and crustaceans, for example—but that the coverage of other groups was too sketchy or scattered to be useful for field work. The first makeshift field guide for students, created and illustrated largely from the literature, provided the impetus to develop a field guide describing and illustrating with color photographs all the common (and many not-so-common) Galápagos seashore invertebrates. Scientists at the Charles Darwin Research Station urged me to begin with the echinoderms, especially the sea cucumbers, which of all echinoderms were the most troublesome to identify, and later to extend the project to other groups.

This ambitious project has necessitated many trips to the Galápagos Islands to collect and photograph invertebrates from the intertidal and subtidal environments throughout the archipelago. The result is this book, in which species descriptions are brought together with original color photographs of the living animals, many of which are presented in published form for the first time.

Acknowledgments

The preparation of a field guide like this requires the assistance and cooperation of many people. William C. Ober has accompanied me on several trips to the Galápagos to help with collecting and underwater photography. Assisted by Claire Garrison, he also painted color illustrations of many of the echinoderms and prepared the line art. A guide book of this kind would be impossible without the support of authorities who made or verified identifications of numerous specimens. David Pawson of the Smithsonian Institution provided assistance with the troublesome sea cucumbers on several occasions. Gordon Hendler of the Los Angeles County Museum of Natural History repeatedly contributed his expertise on brittle stars. Others who read sections and offered suggestions were Chris Mah (sea stars) and Rich Mooi (echinoids), both of the California Academy of Sciences, and Harris Lessios, (echinoids) Smithsonian Tropical Research Institute. Gerard Wellington's 1975 unpublished report to the Department of National Parks and Wildlife, Quito, the result of an extensive two-year study of the Galápagos coastal marine environment, was a rich source of information on the distribution and ecology of many of the species described herein.

It is also a pleasure to acknowledge the cooperative assistance of the personnel of the Charles Darwin Research Station who made arrangements for collecting trips and offered encouragement and support throughout the project. The National Park Service provided the required permits for visiting and collecting specimens at study sites throughout the archipelago. Although many have helped in countless ways, I especially wish to thank Rodrigo Bustamante who, with Priscilla Martinez, helped in getting the field guide through its gestation by making available work space and facilities in the BioMarine laboratory at the Darwin Station, arranged boat charters, provided diving equipment, and assisted in field research—all accompanied by sound advice and cheerful sense of humor. Others who accompanied me on one or more dive trips and helped in the collecting of specimens include George Branch, Robert Day, Jorge Gomezjurado, Scott Henderson, Hugh Jarrard, Shannon Jones, Joshua Nitsche, Jimmy Peñaherrera, Juan Carlos Ricuarte, Will Shepherd, Rita Spadafora, and Robert Van Syoc. Larry Roberts offered detailed helpful advice on the manuscript, and my daughter, Diane Hickman Liss, meticulously proofed the final draft. My wife Rae accompanied me on several trips and exercised commendable patience while I was preoccupied with this project. Finally, I will always be grateful to the many students from Washington and Lee University, whose involvement in an intertidal censusing program begun in 1989 convinced me of the need for a field guide such as this one.

How To Use This Field Guide

This field guide to the Galápagos echinoderms illustrates and describes in some detail the common sea stars, brittle stars, sea urchins, and sea cucumbers that occur in the intertidal zone and in shallow water down to about 30 m (100 feet). Species accounts include descriptions of diagnostic features, habitat, geographic range, depth range, comments on biology of the species, and ways to distinguish similar species. With few exceptions, photographs were taken of living animals, either photographed against a black background in an aquarium, or in their natural habitats. Because some specialized terminology is often required to describe aspects of an animal's morphology, key anatomical features are shown in labeled drawings that appear at the beginning of each chapter.

Most species can be identified by simply thumbing through the book to find the photograph depicting the animal in question. However, since some species closely resemble one another, it is important to read the diagnostic field characters in the description to confirm your identification. Also, be sure to note whether or not the animal's described habitat matches the habitat in which you found the animal. Most species can be identified without using magnification, but it is often necessary to examine the anatomy of brittle stars closely to verify an identification. For this purpose a simple hand lens is handy.

Sea cucumbers present a special challenge for identification. Although many sea cucumbers can be identified by external characteristics such as color patterns and surface features, others are almost impossible to identify with certainty except by examining the microscopic ossicles embedded in the cucumber's body wall. This is an option not available to most readers, but for the benefit of scientists who have access to a compound microscope, I have included in Appendix A descriptions of the ossicles for all the sea cucumbers included in this field guide, together with directions for preparing the body-wall ossicles for examination.

Each species is listed by its two-part Latinized scientific name, called a binomial. The first part of the name is the genus, which is capitalized, and the second part is the species epithet, which is not capitalized. The scientific name is always printed in italics. Following the scientific name is the name of the individual (called the authority) who first described the species, and the date of the original publication. Often on the basis of subsequent research it is decided that a particular species should be assigned to a different genus. When this happens, the original author's name is placed in parentheses.

I've included common names where these are known, but unlike the Galápagos fishes, which bear well-recognized common names in most international languages, the majority of echinoderm species in Galápagos have not been named by laymen. In some instances I have created a common name when characteristics of the animal make a name obvious and fitting. However, because common names are loosely applied, are necessarily restricted to familiar forms, and often include several similar species, the only unmistakable way to refer to a species is by its scientific name. The sea cucumbers have no common names.

Bear in mind that whereas I have tried to include all echinoderm species that are commonly encountered in Galápagos, there are species not described here that you may be fortunate in sighting. Conversely, there are common species characteristically concealed under rocks, especially during the hours of daylight, that only can be encountered by turning over rocks—a disturbance of the habitat that is not allowed in the Galápagos National Park without a scientific permit. Because I developed this field guide to assist scientists in their field work as well as interested visitors to the Galápagos, I have included a listing of selected references that may be consulted for additional information. Appendix C is a guide to the most useful references for each species.

THE PHYLUM ECHINODERMATA

IN GALÁPAGOS

O ne of the delights of visiting the Galápagos seashore is the discovery of sea stars, sea urchins, and sea cucumbers, all curiously beautiful representatives of the phylum Echinodermata. The name Echinodermata (ee-ki′no-der′ma-ta) (Gr. *echinos*, prickly, + *derma*, skin) perhaps applies best to sea urchins, which are armed with movable spines. Echinoderms are a strange group sharply distinguished from all other members of the animal kingdom. The phylum is an ancient group extending back to the Cambrian period, some 600 million years ago. Although the activities of echinoderms seldom resemble those of other kinds of animals, they are more closely related to vertebrates than is any other major group of invertebrates. The echinoderms possess a constellation of characteristics that are unique in the animal kingdom. Despite years of extensive research we are still far from understanding many aspects of echinoderm biology. The distinguished American zoologist Libbie Hyman once described the echinoderms as a "noble group especially designed to puzzle the zoologist."

Echinoderms have successfully exploited radial symmetry, a feature that allows them to engage the environment from any direction. Oddly, despite the radial symmetry of adults, echinoderms are clearly descended from bilateral ancestors, a fact confirmed by their bilateral larvae, which become radially symmetrical later in development. All echinoderms are built on a basic five-part plan, the five parts arranged around a central oral-aboral axis. This plan is best seen in a sea star in which five arms radiate out from the central disc, with the central axis defined by the mouth on the under (oral) side and the region opposite the mouth on the upper (aboral) side of the animal. One of several unique features of echinoderm anatomy is the

water-vascular system that uses hydraulic power to operate a multitude of tiny tube feet used in food gathering and locomotion. The tube feet project from grooves that extend outward on the under side of the arms in radii called ambulacra, a word derived from the Latin meaning "to walk." Sea stars have open ambulacral grooves, but in many echinoderms—brittle stars and sea urchins, for example—the ambulacra are not evident as furrows and the animals are said to have closed ambulacral grooves.

Another odd feature of echinoderms is that in some species the endoskeleton of dermal ossicles or plates may fuse together to invest the animal in armor, while in other species the skeleton may be reduced to microscopic bodies embedded in the body wall. Sea stars and sea urchins possess minute, pincer-like pedicellariae scattered over the body surface that help keep the body surface free of debris, and may aid in food capture. Perhaps strangest of all is the absence in echinoderms of a brain or anything resembling a head. They manage with a nervous system consisting mainly of a nerve ring surrounding the gullet from which extends a system of radial nerves that coordinate movement and report on changes in the animal's surroundings.

Because echinoderms have limited capacity for osmotic regulation, they seldom venture into brackish waters. They are bottom dwellers found living in all oceans at all depths and are often the most common animals in the deep ocean. Approximately 6,500 living species and 13,000 fossil species of the phylum Echinodermata have been described. **Sea stars** (class Asteroidea), often called starfish, have five or more arms radiating from a central disc. Sea stars seem stiff when picked up, but given time they can twist themselves into any position. Their stiffness is due to an endoskeleton of small plates hinged together to allow movement. **Brittle stars** (class Ophiuroidea) have five (sometimes six or more) snakelike arms attached to a small central disc. Their arms may be branched, sometimes extensively. They are the most active of echinoderms, but because they are concealed during the day they are less familiar than sea stars and sea urchins. **Sea urchins** (class Echinoidea) have a compact, globose body lacking arms. The body, or test, is armed with moveable spines of various lengths. Most living species are "regular" urchins of subspherical shape, radial symmetry, and bear medium to long spines. The "irregular" urchins, the heart urchins and sand dollars, are ovoid or discoid in shape and bear short spines. **Sea cucumbers** (class Holothuroidea) have a cylindrical body that lacks arms or spines, is usually soft walled, and has a mouth surrounded by feeding tentacles. Sea cucumbers have evolved bilateral symmetry as adults with right and left sides as well as upper and lower surfaces.

Echinoderms are well represented in Galápagos. According to Maluf (1991) nearly 200 species of echinoderms have been collected from the Galápagos Islands and on oceanic ridges surrounding the Galápagos. However, the majority are deep-water species never encountered in sport diving. Only approximately one-half of the 200 Galápagos species are found at depths above 200 m (650 feet) and less than 20% of these are found exclusively above 20 m (65 feet). Approximately 17% of Galápagos species are endemic, that is, found exclusively in the Galápagos Islands.

The vast majority of echinoderms live in the subtidal zone, the area below the intertidal zone that is always submerged. Some, however, inhabit the intertidal zone, also called the littoral zone, where sea and land meet. This is one of the harshest of all marine environments, subjected to pounding surf, sun, wind, rain, extreme temperature changes, and sedimentation. Nevertheless, beneath the black lava rocks of Galápagos shores there dwells a rich diversity of life, including some of the hardier sea urchins, brittle stars, and sea cucumbers. Most inhabit the low littoral zone which is exposed to air only briefly during low tide. Exposure time increases the higher one goes in the intertidal zone; the upper edge of the high littoral zone, in Galápagos about 2 meters above the low tide mark, may be submerged only during the highest tides. Above this is the spray zone which is never submerged but is wetted by sea water spray.

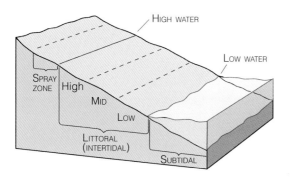

Typical zonation pattern on a Galápagos beach. Animals living in the intertidal (littoral) zone are profoundly influenced by the rise and fall of the tides and by exposure to wave action. In Galápagos the average tidal range is about 2 meters.

4

Sea Stars

CLASS ASTEROIDEA

*S*ea stars are the "prima donnas" of echinoderms, familiar to many people as beautifully symmetrical symbols of marine life. In Galápagos, however, sea stars (often called starfish) are much less common than sea urchins and brittle stars. Like all echinoderms, they are strictly bottom dwellers. Despite their striking coloration and large size that make them conspicuous on the sea bottom, they seem to have few enemies, suggesting that sea stars produce something that repels would-be predators. They are diverse feeders. Some sea stars are particle feeders but most are predators of sedentary or sessile prey since sea stars are themselves slow-moving animals. Carnivorous forms prey on molluscs (their favorite food), crabs, corals, worms, or other echinoderms. Others are scavengers that feed on decaying fish and invertebrates. Still others are deposit feeders, filling their stomachs with mud from which they extract organic material.

Forty-four sea stars are reported from the Galápagos Islands, most of which are deep water species that are collected by dredging. Of the 14 species reported from the intertidal and shallow water, only six or seven are at all common. Most sea stars occurring in Galápagos are of tropical East Pacific distribution; only two (possibly three) of the shallow-water sea stars are endemic to Galápagos.

Sea Star Body Plan

Sea stars typically have 5 arms (sometimes more) radiating from a central disc that bears the mouth on the lower (oral) surface and an inconspicuous anus on the upper (aboral) surface. A circular, sieve-like madreporite, also on the upper surface, leads to the water vascular system. The upper surface of most sea stars is covered with blunt or sharp spines, although in some the spines are flattened so that the surface appears smooth. Around the bases of the spines of many sea stars are small pincer-like structures, the pedicellariae, that are used to keep the body surface free of debris and prevent marine larvae from settling there. The surface is also provided with small, delicate, retractile projections, the dermal gills, which function in respiration. The surface of some sea stars is covered with paxillae, tiny skeletal columns with brushlike crowns. When viewed from above with magnification, the paxillae look like a field of flowers. On the under side of the sea star long ambulacral grooves radiate out along the arms from the mouth; from these grooves project two rows of tube feet.

The size of sea stars is expressed as the radius—the distance between the center of the disc to the tip of an arm—for an average-sized star.

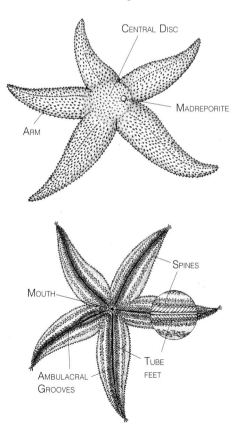

External anatomy of a sea star, aboral (dorsal) view.

External anatomy of a sea star, oral (ventral) view.

6

Family Astropectinidae

Astropecten armatus Gray, 1840   **Spiny Sand Star**

Radius 8 cm (3.2 in)
This handsome, medium-sized species varies in color from bright red to rich brown. It has a flattened body with long arms, wide at the base, and tapering to the ends. The aboral (upper) surface is pebbled with paxillae and the arms are bordered by sharp spines. The tube feet lack suckers. Although its feeding behavior in Galápagos has not been studied, in California *A. armatus* preys mainly on snails (which it swallows whole), dead fish, and sand dollars. **Habitat & range:** On sandy or crushed coral bottoms, often partly covered by sediments. Southern California to Ecuador and Galápagos Islands.

Family Luidiidae

Luidia foliolata Grube, 1866   **Sand Star**

Average arm radius 8 cm (3.2 in)
The arms of this five-rayed sea star are somewhat constricted at the base, then taper evenly to a sharp point. The paxillae on the aboral surface are square, forming four regular series at the margin of the ray, then decreasing in size toward the median area of the ray where the paxillae are small, irregular, and not arranged in a series. There are three rows of sharp spines at the edges of the rays. The tube feet on the oral surface are very large. Color, dark olive. This species resembles the much larger *Luidia superba*, which also occurs in Galápagos, but is distinguished from the latter by the absence of pedicellariae and spines on the aboral surface. **Habitat & range:** On sand bottoms, subtidal from 3 m (10 ft) or less to dredging depths. This species is especially active at night, when they may be seen moving rapidly across open sandy surfaces, such as the Tagus Cove anchorage where a large population thrives. Eastern Pacific coast from Alaska to southern Mexico, and the Galápagos Islands.

Luidia bellonae Lütken, 1864 — **Banded Sand Star**

Radius 10 cm (4 in)
This star has five slightly convex arms bordered by a fringe of sharp, marginal spines. The tube feet lack suckers. The upper surface is covered with paxillae, irregularly arranged in zones in the center of the arms, but toward the sides of the arms they become larger and disposed in longitudinal rows. The crowns of the paxillae consist of tiny spines clustered together. Body color is yellowish to lavender with brown spots on the dorsal surface, these spots arranged more or less in transverse bands.
Habitat & range: Subtidal sandy bottoms. They lie buried just beneath the surface where the star-shaped outline of the body is often visible. More numerous in the western archipelago but nowhere common. Gulf of California to Peru and Galápagos Islands.

DRIED SPECIMEN, CHARLES DARWIN RESEARCH STATION

Family Oreasteridae

Nidorellia armata (Gray, 1840) — **Chocolate Chip Sea Star**

Radius 8.5 cm (3.3 in)
This distinctive and easily recognizable sea star has blunt arms broadly attached to a large central disc, giving it an inflated appearance. The large aboral spines bear dark tips which stand out against a tan ground color. There is considerable variation in form and color throughout the archipelago.
Habitat & range: Rocky shore, intertidal to 73 m (240 ft); common throughout Galápagos. Feeds on benthic algae, small gastropods, and sessile invertebrates such as tubeworms. Gulf of California to Peru and Galápagos Islands, also Hawaii.

Pentaceraster cumingi (Gray, 1840)   **Panamic Cushion Star**

Radius 17 cm (6.7 in)
The upper surface of this large sea star is covered with blunt, immobile spines. The dorsal plates and spines are bright red, the spaces between greenish-brown, but there is considerable variation in color. This species feeds on microorganisms on sea grass and algal substrates. It also may feed on sea urchins and other echinoderms by extraoral feeding, that is, by everting the stomach to cover and digest the prey, then absorbing the nutrients.
Habitat & range: Sandy bottoms, extreme low intertidal to 180 m (590 ft); seldom seen above 4 m (13 ft) depth. Often quite numerous at common snorkeling and dive sites in the central archipelago. When viewed from the surface this sea star appears black in color because of the absorption of red light in sea water. Gulf of California to Peru and Galápagos Islands, also Hawaii.

Family Asterinidae

Asterina sp., cf. *Asterina modesta* Verrill, 1870   **Blunt-rayed Sea Star**

Radius 1.5 cm (0.6 in)
This very small, 5-rayed star bears extremely blunt, broad rays which protrude only slightly from the elevated central disc. No protruding spines. Color variable with tan or brown predominating. Because this genus is in need of thorough revision, the species of the Galápagos representative is uncertain. The genus contains about 30 species of cosmopolitan distribution; all are confined to the littoral zone.
Habitat & range: Under rocks in intertidal zone of sheltered areas, common but not easily seen because of small size and habit of clinging securely to underside of stones. *A. chilensis* is reported from Peru and Chile; the range of the Galápagos species is not known.

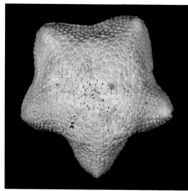

Dried specimen, Charles Darwin Research Station

Family Asteropseidae

Asteropsis carinifera (Lamarck, 1816)   **Keeled Sea Star**

Radius 16 cm (6.3 in)
The arms of this large sea star are acutely triangular in cross section, with a mid-radial ridge emphasized by a series of spines and pores in small groups. The body is covered with a thick, smooth skin, giving the animal a wet or slimy appearance. The oral side is flat, the aboral side very convex. The edges of the arms are armed with prominent conical spines. Coloration is tan with brick red pigment splotches on the central disc and arms.

Habitat & range: Rocky reefs, coral rubble, and underside of large rocks and coral slabs. Intertidal to about 35 m (115 ft). Uncommon except in certain areas such as Rabida Island. Throughout the Indo-Pacific and from the Gulf of California to Panama and Galápagos.

Family Ophidiasteridae

Linckia columbiae Gray, 1840   **Variable Sea Star**

Radius 3.5 cm (1.4 in)
This is a small, narrow-rayed sea star normally with five arms but often with fewer or more than five. The central disc is small, the arms are rounded, and the upper surface of arms and disc is leathery and pitted and without spines or pedicellaria. Coloration is variable, from mottled red or orange-brown to tan and purple. This species is variable in appearance and is often found with one or more missing or partly regenerated arms. It is a ciliary feeder that ingests the film of microorganisms adhering to hard surfaces.
Habitat & range: Low rocky intertidal under stones and in crevices, and subtidal to 150 m (490 ft). More common in the western archipelago. Southern California to Peru and Galápagos Islands.

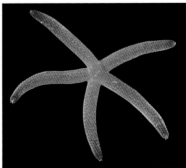

"Comet" form of *L. columbiae* (*above*), formed from asexual fission of adults (*right*) that voluntarily detached their arms. Arm stubs will grow complete arms.

Leiaster teres (Verrill, 1871)   **Smooth Sea Star**

Radius 20 cm (8 in)
This large, brilliant red sea star is distinguished by its shiny, smooth skin, small disc, and long, rounded arms. There are three dorsal rows and two lateral rows of overlapping scales on the arms.
Habitat & range: Rocky and mud/sand bottoms from shallow water to about 50 m (165 ft). Uncommon. Gulf of California to Panama and Galápagos Islands (new record).

Pharia pyramidata (Gray, 1840)   **Pyramid Sea Star**

Radius 12 cm (4.7 in)
This medium-sized sea star has five rather robust arms, triangular in cross section. The central disc, which projects above the rays, bears a large, irregular, composite madreporite in the angle between two arms. There are eight rows of dermal branchiae (4 dorsal, 2 lateral, 2 oral) which tend to join toward the sides. Color purplish brown to dull orange-brown. This species can be distinguished from *Phataria unifascialis* (which it resembles in shape) by its wider ambulacral grooves lined with a single row of stubby, flattened spines (*Phataria* has a double row) and by the triangular shape of the arms.
Habitat & range: Low intertidal and subtidal rocks to 130 m (425 ft). Locally common in Galápagos; often seen by snorkelers at Devil's Crown. Gulf of California to Peru and Galápagos Islands.

Closeup of central disc area.

Phataria unifascialis (Gray, 1840)　　　**Blue Sea Star**

Radius 9 cm (3.5 in)
The body of this handsome sea star is slender and stiff, with a small disc and long, tapering arms with a smoothly pebbled surface. Coloration is variable but usually the aboral (upper) surface is royal blue bordered with dark blue or black bands running the length of the rays. Oral surface is orange-red along ambulacral grooves. There is a double row of blunt plates lining each side of the ambulacral grooves. Feeds on algae.

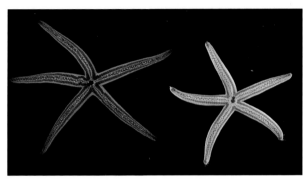

Habitat & range: Rocky shores low intertidal to subtidal to about 50 m (165 ft). Snorkelers often see this star at Devil's Crown. Gulf of California to Peru and Galápagos Islands.

Two color morphs of *Phataria unifascialis.*

Family Mithrodiidae

Mithrodia bradleyi Verrill, 1867　　　**Bradley's Sea Star**

Radius 18 cm (7.1 in)
This large red sea star has subcylindrical arms somewhat wider than thick and partially constricted where united with the small central disc. At the sides of the arms is a row of irregularly spaced thick, blunt spines; large spines are also scattered irregularly on the dorsal surface between the lateral rows of spines. This star will shed its arms (called autotomizing) if handled roughly.
Habitat & range: Intertidal rocky to about 50 m (165 ft). Locally common in Galápagos. Central Gulf of California to central Pacific including Galápagos Islands.

12

Family Acanthasteridae

Acanthaster planci (Linnaeus, 1758) **Crown-of-Thorns**

Radius 10-15 cm (4-6 in)

The multi-rayed crown-of-thorns is easily distinguished from any other Galápagos sea star by its pink to orange color and the numerous long, sharp (and toxic), white spines that cover the upper surface of the 10 to 14 arms and central disc. The star blanches quickly when touched. Although the ecology of

this species has not been studied in Galapagos, casual observations suggest that it is an important predator of coral in Galapagos as it is elsewhere in its range. **Habitat & range:** Subtidal to about 20 m (65 ft), especially in rocky areas of coral and gorgonian growth. Rare in Galápagos. It was first sighted at the northern island of Darwin in 1995 and is frequently seen at the Darwin anchorage. Distributed throughout the Indo-Pacific on coral reefs.

Family Heliasteridae

Heliaster cumingii (Gray, 1840) **Red Sun Star**

Radius 9 cm (3.5 in)

This sea star has 32 to 40 rays (average 36) with only about 25% of the ray free; the remaining 75% is fused to adjacent rays. The central disc is large. Color on the dorsal surface varies from red to deep bluish-black.

Habitat & range: Rocky shores, closely associated with the mid-littoral *Tetraclita* barnacle zone. This species was common in Galápagos before the 1982-83 El Niño event, but now is rare. Both species of *Heliaster* feed on barnacles and other sessile invertebrates. Endemic to Galápagos.

Heliaster solaris (A.H. Clark, 1920)   **Twenty-four-rayed Sun Star**

Radius 14 cm (5.5 in)

This sunstar bears 21 to 27 rays (average about 24). About 60% of each ray is free, with the remaining 40% fused to adjacent rays. The central disc is abruptly elevated at the center. The dorsal surface is light gray or greenish-yellowish in color, irregularly blotched with dark gray or black. On the rays, the dark blotches look like irregular cross bands.

Habitat & range: Rocky shores, midlittoral but distributed somewhat lower on the beach than *H. cumingii.* This species was uncommon before the 1982-83 El Niño, apparently has not been sighted since. Endemic to Galápagos.

DRIED SPECIMEN, CHARLES DARWIN RESEARCH STATION

Other Sea Stars
Reported from the Galápagos Islands

The following sea stars have also been reported from the Galápagos intertidal and shallow water habitats, but are uncommon or rare: *Luidia superba* A.H. Clark, 1917; *Luidia columbia* (Gray, 1840); *Astropecten fragilis* Verrill, 1867; *Narcissia gracilis* A.H. Clark, 1916; *Astrometis sertulifera* (Xantus, 1860); and *Paulia horrida* Gray, 1840.

Brittle Stars

*B*rittle stars are the most abundant echinoderms in Galápagos, yet they are seldom seen by the casual observer. They are secretive animals that hide under and between intertidal and subtidal rocks by day to escape predation by fish. Even at night they may cautiously extend only their arms from hiding places to feed. Brittle stars are the most agile of echinoderms and if disturbed (for example, by lifting the rock under which they are hiding) they move rapidly, using quick rowing movements of the arms to scuttle to safety.

Brittle stars feed on a variety of small organic particles suspended in the water or lying on the bottom. Food is trapped on mucous strands and transferred to the mouth by the podia (tube feet) or by coils of the arms. They also feed by absorbing dissolved compounds through their skin. Brittle stars have the notorious ability to break off their arms, a defensive response to attack by a predator. It is common to find individuals with missing or partly regenerated arms.

Of the approximately 2000 described species of brittle stars, 74 have been recorded from Galápagos, 28 of which occur in water less than 20 m (65 feet) deep. Of the 12 species listed here, four are believed to be endemic to the Galápagos Islands.

Brittle Star Body Plan

The arms of brittle stars are slender, sharply set off from the central disc, and lack the pedicellariae and dermal gills characteristic of sea stars. The jointed arms consist of articulated ossicles (called "vertebrae") connected by muscles and covered with dermal plates. The ambulacra are closed in brittle stars. The tube feet, important in locomotion and in obtaining food, are small, lack suckers, and project laterally from the skeletal plates.

Measurements of disc diameter and arm length, given in metric units and their U.S. equivalents, refer to average dimensions of adult specimens.

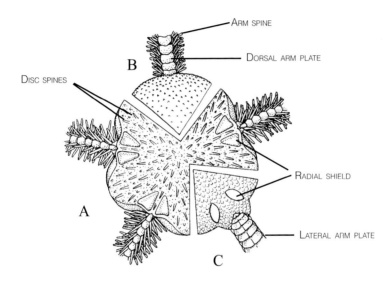

Composite external anatomy of the dorsal surface of the disc and arms of brittle stars, with sectors showing the typical features of A, *Ophiothrix spiculata*; B, *Ophiocoma alexandri*; and C, *Ophioderma teres.*

16

Family Amphiuridae

Ophiophragmus sp. Lyman

Disc 4.3 mm (0.17 in), arm length 65 mm (2.5 in)
This is a small, five-armed brittle star with arm length nearly 10 times the diameter of the disc. The long, delicate arms readily distinguish this genus from others occurring in Galápagos. Four species of this genus have been recorded from Galápagos (Maluf, 1991). **Habitat & range:** Subtidal under rocks. Of the four species recorded from Galápagos, one (*O. disacanthus*) is endemic to Galápagos and the other three range from southern California to Galápagos.

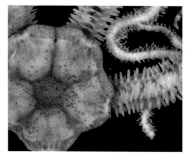

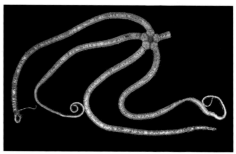

Detail of disc and arms.

The long arms of *Ophiophragmus* distinguish this genus.

Family Ophiactidae

Ophiactis simplex (Le Conte, 1851)

Disc 2 mm (0.08 in), arm length 6 mm (0.24 in)
This very small, five- or six-armed brittle star is very similar to *Ophiactis savignyi* in appearance but differs from the latter in small ways, one of which is having much smaller, almost indiscernible, radial shields. Broad dorsal arm plates are characteristic of this species. It multiplies asexually by dividing across the disc with each half then regenerating the missing parts (see photograph); it also reproduces sexually. Uncommon in Galápagos.
Habitat & range: Rocky shores, from mid to low littoral down to dredging depths. It is associated with sponges, tube snails, and clumps of coralline and other algae. Southern California to Peru, Cocos, and Galápagos Islands.

This specimen of *Ophiactis* has recently divided asexually across the disc and is regenerating three new arms.

Ophiactis savignyi (Müller & Troschel, 1842)

Disc 2.8 mm (0.11 in), arm length 11.5 mm (0.45 in)
This is a small six-armed brittle star distinguished by the large, brown pigmented radial shields, their lengths typically exceeding half the radius of the disc. There are five to six arm spines of about the same size on each side of an arm joint; the spines are blunt, about as wide at the tip as at the base, and have many little spines at the tip. Small spines cover the disc, mainly in the interradial areas and edges. The aboral (dorsal) arm plates are about twice as wide as long. The color is variable with various combinations of greenish yellow, brown, and cream. There is usually a white patch at the outer tip of the darkly pigmented radial shield and pairs of white spots at the outer edge of the dorsal arm plates. This species reproduces both sexually and asexually (by fission across the disc). **Habitat & range:** Intertidal and shallow water to dredging depths, among coral, coral rubble, algae, and sponges. Cosmopolitan in tropical and subtropical regions.

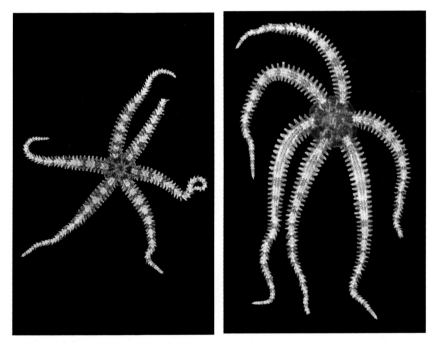

Although these two specimens of *Ophiactis savignyi* show variation in pigmentation, the characteristic brown radial shields are evident in both.

18

Family Ophiotrichidae

Ophiothrix spiculata LeConte, 1851   **Glass-spined Brittle Star**

Disc 14 mm (0.55 in), arm length 64 mm (2.5 in)

This small brittle star, the most commonly seen brittle star in Galápagos, both intertidally and subtidally, is easily recognized by the long, glasslike spines that extend laterally from the arms. The spines are blunt at the tip, flattened, about as wide at the tip as at the base, and bear distinct, thornlike teeth. The smallest spines of the disc are multifid (ends split into two or three minute spinelets). This last character, which requires a hand lens to be seen, is crucial in separating this species from its close relative *Ophiothrix magnifica. O. magnifica,* which lacks multifid spines on the disc, occurs in Peru and probably in Galápagos, but all the specimens of *Ophiothrix* that I have examined belong to the ubiquitous species *O. spiculata*. According to ophiuroid specialist Gordon Hendler, specimens resembling *O. magnifica* may be variants of *O. spiculata*; the taxonomy requires clarification. Coloration is highly variable, ranging from blue or gray-green to violet, arms nearly always banded at intervals. Intertidal specimens tend toward dark, cryptic colors; subtidal specimens are often brightly colored. This species feeds by capturing food on sticky secretions of the arm spines and tube feet.

Habitat & range: Very common mid and low intertidal under rocks and on large sponges, or clinging to algae; also common subtidally under rocks. Often abundant among roots of red mangrove in association with estuarine sponges. California to Peru and the Galápagos Islands.

O. spiculata from an intertidal habitat.

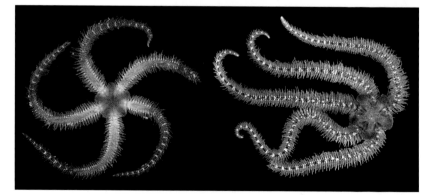

Two color morphs of *O. spiculata* from a subtidal habitat where coloration tends to be more vivid. This most common of Galapagos brittle stars is variable in coloration.

Ophiothela mirabilis (Verrill, 1867)   **Epizoic Brittle Star**

Disc 2.4 mm (0.1 in), arm length 12.5 mm (0.5 in)
This tiny, six-armed brittle star is most easily identified by its habitat. It is epizoic, clinging tightly to the spines of the pencil urchin (*Eucidaris thouarsii*) or the branches of gorgonians. The disc, ranging in size to 5 mm, but seldom larger than 2 mm in the Galápagos specimens examined, is mostly covered with large radial shields which may reach nearly to the center of the disc and are narrow and blunt at their outer ends. The lateral arm plates bear five or six (usually six) spines. The uppermost spine is very small, the second or third spines are the longest, and the others decrease in size to the lowest, which is very small. It reproduces asexually by dividing across the disc, so it is common to find specimens with partly regenerated arms. The disc may be distorted, wide at the base of the long arms, and compressed at the bases of the short, regenerating arms. Species of *Ophioactis* also are sometimes epizoic and could be confused with *O. mirabilis*.
Habitat & range: Epizoic on pencil urchins and gorgonians, and possibly on sponges. It has been reported from the Pearl Islands, Bay of Panama, Malpelo Island, and now from the Galápagos Islands (new record).

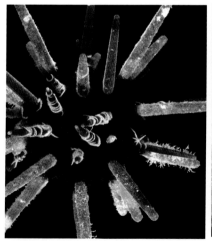

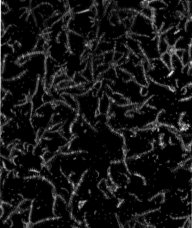

Several *Ophiothela mirabilis* on the spines of a pencil urchin. The epizoic brittle stars keep the spines clear of marine life that would settle there.

Numerous *O. mirabilis* adhere tightly to the branches of a gorgonian. The brittle stars do not harm the gorgonian but use it as a feeding station to trap plankton passing by in the current.

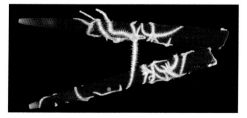

Two spines removed from a pencil urchin with epizoic brittle stars tightly wrapped around them.

20

Ophiocoma aethiops Lütken, 1859   **Black Spiny Brittle Star**

Disc 3.4 cm (1.34 in), arm length 19 cm (7.5 in)
This robust, large, and very common brittle star is dark brown to black in color.
The disc is covered with fine, round granules. Ventral arm plates are slightly
wider than long, with rounded outer margin. Each lateral arm plate bears four
spines. Unlike *Ophiocoma alexandri* there are alternating numbers of arm spines
on adjacent arm segments. This impressive giant of brittle stars can hardly be
confused with any other species. Specimens with arm lengths much longer than
19 cm have been collected in Galápagos.
Habitat & range: Common everywhere in Galápagos on rocky shores, under
rocks in sand or muddy sand of the lower intertidal to about 30 m (100 ft). Gulf
of California to Panama and Galápagos Islands.

Detail of arm.

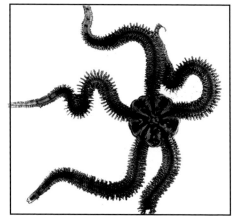

Large adult specimen of *Ophiocoma aethiops.*

An impressive
specimen of *O.
aethiops.* The black
spiny brittle star is the
largest Galápagos
brittle star.

Ophiocoma alexandri Lyman, 1860   **Alexander's Brittle Star**

Disc 2.3 cm (0.9 in), arm length 18 cm (7.1 in)

This large brittle star is similar to *O. aethiops* in appearance but is distinguished by alternate dark gray and white bands on the arms. It is more delicate and more slender-armed than *O. aethiops*. The lateral arm plates bear six spines (sometimes five or seven). The oral shields are almost round. It is common at the same sites where *O. aethiops* is found. Both this species and *O. aethiops* feed on organic particles suspended in the water or lying on the bottom. They forage from dusk to dawn by extending their arms from crevices or from beneath stones to wave the arms about or hold them aloft to capture organic detritus.

Habitat & range: Intertidal to 70 m (230 ft). Common. Southern California to Galápagos Islands.

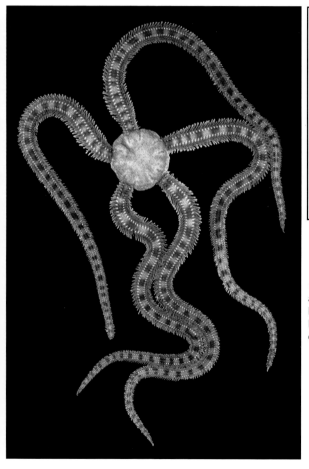

Detail of arm.

Ophiocoma alexandri is a handsomely patterned brittle star, nearly as large as *O. aethiops*, but of more delicate build.

22

Ophiocomella schmitti A.H. Clark, 1939   **Schmitt's Brittle Star**

Disc 5 mm (0.1 in), arm length 14.5 mm (0.57 in)

The disc of this small six-armed species is black, and covered with granules and stout spinelets. The arms bear smooth, stout spines; banding on the arms alternates between black and dark to pale gray. Dorsal arm plates are triangular, wider distally than proximally; lateral arm plates indistinct. This species, like *Ophiactis* spp., divides asexually by fission across the disc.

Habitat & range: Primarily intertidal. Endemic to Galápagos Islands.

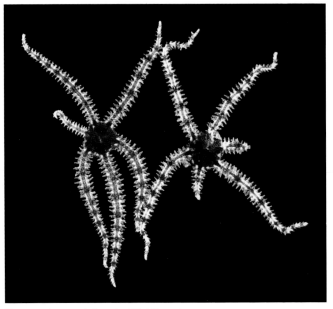

Two specimens of *O. schmitti*. Although relatively common in the intertidal zone, this species is easily overlooked because of its small size.

Detail of arm.

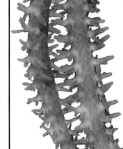

Family Ophioneredidae

Ophionereis albomaculata E.A. Smith, 1877   **White-banded Brittle Star**

Disc 11 mm (0.4 in), arm length 72 mm (2.8 in)

The disc of this handsome, easily recognizable brittle star is pale gray on the upper surface which is covered by skin, thus obscuring fine scales beneath. The arms are black with white bands at intervals. The lateral arm plates bear four, sometimes five, spines.

Habitat & range: Intertidal. Relatively common. This species may be endemic to Galápagos.

Detail of arms and central disc.   *Ophionereis albomaculata* in the intertidal habitat.

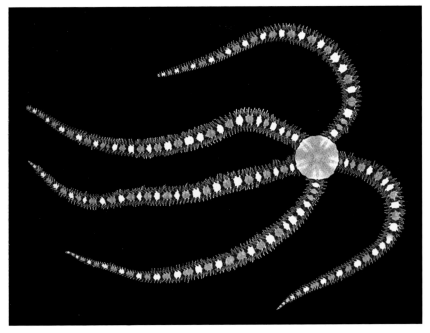

A specimen of *O. albomaculata* collected subtidally.

24

Ophionereis perplexa Ziesenhenne, 1940   **Three-spined Brittle Star**

Disc 7 mm (0.28 in), arm length 31 mm (1.2 in)

This small brittle star has a flattened disc, depressed in the center, and three large, heavy, blunt-tipped arm spines on each lateral arm plate. The dorsal arm plates are wider than long, more or less triangular with convex proximal margin and convex distal margin. The disc appears smooth (numerous overlapping scales visible at high magnification). The radial shields are narrow with edges mostly obscured by minute scales. The arms are banded brown with white borders and the center of the disc is irregularly pigmented brown.

Habitat & range: Intertidal and offshore to dredging depths, under rocks and rubble. Endemic to Galápagos.

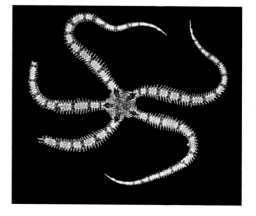

Family Ophiodermatidae

Ophioderma teres (Lyman, 1860)   **Smooth Brittle Star**

Disc 2.2 cm (0.87 in), arm length 6.4 cm (2.5 in)

This is a large, solid, brown to slate-colored brittle star with short arm spines that lie flat against the arms. It is principally carnivorous, feeding on small crustaceans, worms, sponges, and algae, but also a scavenger. The dorsal arm plates of *O. teres* are fragmented into numerous smaller plates, unlike an uncommon relative of this species, *Ophioderma panamense*, which lacks fragmentation of the dorsal arm plates. *O. panamense* also differs from *O. teres* in having light and dark bands on the arms, and paired white spots on the arm tips.

Habitat & range: Under large stones from mid littoral to 45 m (150 ft). Rather common in Galápagos. Gulf of California to Panama and Galápagos Islands.

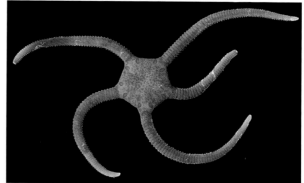

Family Ophiuridae
Ophioplocus hancocki Ziesenhenne, 1935.   Hancock's Brittle Star

Disc 20 mm (0.8 in), arm length 65 mm (2.6 in)
This brittle star has a large disc covered with large and small coarse scales. The disc is large relative to the lengths of the arms. The arm segments are wider than long, have short, blunt spines on the lateral plates, and bear a smooth dorsal pavement of numerous tiny plates of diverse sizes. The color of the disc is cream to tan with irregular black or brown markings, and the arms are banded with black or brown at every sixth segment.
Habitat & range: Tide pools and subtidal under rocks or in sand. Probably endemic to the Galápagos Islands.

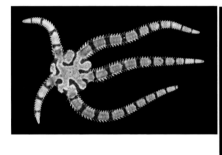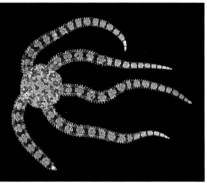

Other Brittle Stars
Reported from the Galápagos Islands

The following brittle stars have also been reported from the Galápagos intertidal and shallow-water habitats. They are considered uncommon or rare: *Amphipholis platydisca* Nielsen, 1932; *Ophiophragmus marginatus* (Lütken, 1859); *Ophiophragmus paucispinus* Nielsen, 1932; *Ophiophragmus tabogensis* Nielsen, 1932; *Ophionereis annulata* (LeConte, 1851); *Ophioderma pentacantha* H.L. Clark, 1917; *Ophioderma variegatum* Lütken, 1856.

Sea Urchins
and Sand Dollars

*A*nyone who has watched a hatpin sea urchin (Diadema *spp.) moving slowly along the bottom can have little doubt about the protective function of the long, sharp spines. The spines are in constant motion. Even a passing shadow causes the urchin to turn more spines in that direction. Other sea urchins have less formidable spines than the hatpin urchin, but all bear the spiny armament suggesting* echinos, *the Greek word for hedgehog, from which the echinoids get their scientific name. Sea urchins move along the bottom or on rock faces by their slender tube feet, which also serve as feelers to detect changes in their surroundings. Many urchins also "walk" along on their spines.*

Zoologists divide the many different kinds of echinoids into two groups based on shape of the test, or "shell," which encloses the body of the urchin. The "regular" echinoids are the sea urchins: radially symmetrical, globose in shape, and armed with long spines. Heart urchins and sand dollars are "irregular" echinoids: they have become bilaterally symmetrical, have a variably-shaped body (flattened in sand dollars), and short spines. In irregular urchins the anus has moved from the aboral pole to a posterior position, and the mouth has moved anteriorly, giving these animals anterior and posterior ends. Irregular urchins always move forward (anteriorly), never sideways as do sea stars, brittle stars, and regular sea urchins.

Sea urchins are herbivorous. Some scrape incrusting algae from rock surfaces, others nibble on plants, and still others trap and eat drifting food. When their regular food is scarce, some urchins will ingest incrusting animals and living coral polyps. Most irregular urchins—heart urchins and sand dollars—feed on sediment particles as they burrow through the upper levels of the sea bottom. The food is moved to the mouth by short spines and tube feet.

Approximately 900 species of echinoids have been described, and are widely distributed in all seas, from the intertidal to deep sea. Thirty-seven species of echinoids have been recorded from the Galápagos. Thirty one occur in water less than 200 m (650 ft) deep, and 25 in water less than 20 m (65 ft) deep (Maluf, 1991). Seventeen of the most common echinoids, the ones you are most apt to see, are described in this guide. None of the common Galápagos sea urchins is endemic.

28

Body Plan of Sea Urchins and Sand Dollars

Sea urchins and sand dollars have a compact body enclosed in an endoskeleton, called the test, composed of closely fitting plates that cannot be moved. As the sea urchin grows, new plates are added to the test from a circle of plates on the dorsal surface. Although lacking arms, the test of echinoids reflects the five-part plan of all echinoderms. This is best seen on an empty, cleaned test, which bears the five rows of little holes from which the numerous tube feet extend in life. In irregular urchins many of the tube feet are arranged into "petals" on the upper surface (see pp. 35-38 for examples). The spines of sea urchins are movable; each spine is attached to the test by a ball-and-socket joint and its movement is controlled by a tiny muscular collar. The socket of each spine fits over a rounded knob (tubercle); these knobs also are best seen on the test of a dead urchin. Also present on the test are stalked, pincerlike pedicellariae. These function to discourage predators and pick off unwanted larvae. Especially active in defense are the toxic, globiferous pedicellariae of some sea urchins, such as the flower sea urchin of Galápagos (p. 33). On the oral or lower side of the urchin is the mouth which, in most sea urchins and sand dollars, is equipped with five teeth that are operated by a complex chewing mechanism called Aristotle's Lantern.

Measurements of overall diameter given in the species descriptions represent horizontal distance between opposite spine tips of adult specimens.

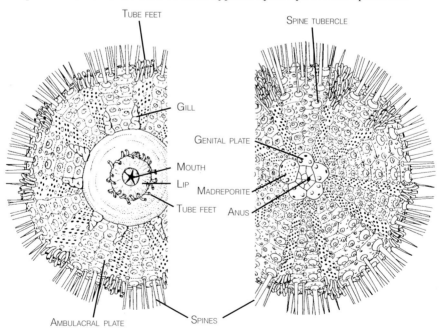

External structure of a sea urchin. The oral (*left*) and aboral (*right*) surfaces are shown with the spines partly removed.

Family Cidaridae

Eucidaris thouarsii (Valenciennes, 1846)   **Slate Pencil Urchin**

Overall diameter 13 cm (5.1 in)

This common urchin is distinguished by ten vertical rows of five to eight clubbed spines each. The spines vary in shape from short and truncate to long and tapering with a chisel-like tip. The spines are often encrusted with sponges, bryozoans, and calcareous algae. This species is considered an indicator species (along with the purple urchin *E. vanbrunti*) of the infralittoral zone. Clark (1948) notes that the "largest and finest specimens" of this species are from the

Galápagos Islands. The slate pencil urchin grazes on encrusting algae and, in Galápagos, is an important predator on *Pocillipora* and *Pavona* coral.

Habitat & range: Very common along Galápagos shores, from low littoral to 20 m (65 ft) and reported elsewhere in its range to 150 m (490 ft). Upper Gulf of California to Ecuador and Galápagos Islands.

Family Diadematidae

Astropyga pulvinata (Lamarck, 1816)   **Cushion Urchin**

Overall diameter 40 cm (16 in)

This handsome urchin has a somewhat flattened, flexible, reddish-brown test with five, conspicuous triangular interradial spots. There is a bluish, inflated anal cone in the center. Young individuals are more brightly colored than adults. The long, fragile spines contain a toxin known to produce localized paralysis from puncture wounds.

Habitat & range: Little is known of the habitat preferences of this uncommon species. In Galápagos it has been sighted only on the northeast side of Albany Island (near Buccaneer Cove on Santiago Island) at about 12 m (40 ft) depth

where they aggregate in large groups numbering 2-300 individuals. The groups disappear in November, presumably migrating into deeper water, and return in February. Outside of Galápagos it has been observed on rock, sand, and mud substrates. Gulf of California to Ecuador and the Galápagos Islands.

Centrostephanus coronatus (Verrill, 1867)   **Crowned Sea Urchin**

Overall diameter 25 cm (9.9 in)

This is an attractive brown urchin with long, acute spines that are distinctly banded with alternating dark and light purple (sometimes almost white) pigment bands; banding is more distinct in small specimens. The uppermost interambulacral plates bear short spines with bright reddish purple tips. The five pairs of oral plates surrounding the mouth bear slender, blunt, light-colored spinelets. The primary spines are not poisonous but are very sharp and brittle and easily penetrate the skin.

Habitat & range: Rock and sand sheltered substrates, usually found in round depressions in rocks that it may excavate. Common. Low intertidal to 125 m (410 ft). Gulf of California to northern Peru and the Galápagos.

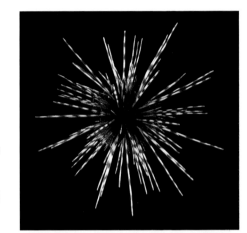

Robert Van Syoc

Two crowned sea urchins living in recesses hollowed in a lava rock wall.

Diadema mexicanum A. Agassiz, 1863   **Hatpin Urchin**

Overall diameter 50 cm (20 in)

The hatpin urchin is a large, black or deep purple sea urchin with very long, acutely pointed, poisonous spines. The primary spines may exceed twice the diameter of the test and are very fragile. Spines are absent from the area around the peristome. In very young specimens the spines are banded with white, but this is replaced with black or dark brown pigment as the urchin ages. Adults are virtually uniformly black or very dark brown, although the spines may be encrusted with white bryozoan colonies. The spines are covered with a thin poisonous membrane that produces somewhat painful, although seldom dangerous, wounds. (The use of vinegar and antihistamines are standard treatments for spine punctures.) This species, along with the other two diadematids, is primarily herbivorous but also feeds on animals and occasionally on coral.

Habitat & range: Rocky low intertidal (tide pools) and subtidal in rocky recesses; a nocturnal feeder. Spotty distribution in Galápagos, occasionally abundant. Mid-Gulf of California to Colombia and Galápagos Islands.

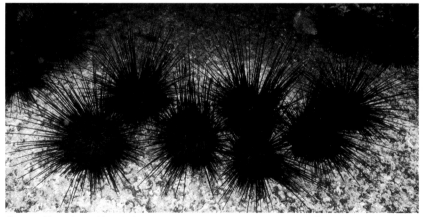

Group of hat-pin urchins clustered in a community at Beagle Rocks. Such clusters are common throughout the archipelago.

Family Toxopneustidae

Lytechinus semituberculatus (Agassiz & Desor, 1846)   **Green Sea Urchin**

Overall diameter 8 cm (3.2 in)
The primary spines of this short-spined urchin are yellow-green. The test of young specimens is deep green above, pure white beneath, with purplish-red patches on the dorsal interambulacral areas. The green urchin is common in tide pools and shallow water. It often displays a "covering reaction" by shielding itself with pieces of shell or bits of algae, held in place by its tube feet. The covering reaction may be a protective response to the intensity of sunlight, for camouflage, or for stability; its purpose remains unclear. The green urchin is a

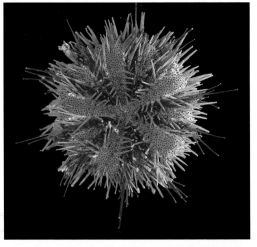

browser that nibbles on plants and debris with its five, sharp, chisel-like teeth.
Habitat & range: Tide pool and lower littoral where it is often abundant, and down to 134 m (440 ft). Very common. It is associated with the rocky habitat near sand patches, in areas of low wave impact. Columbia to Peru and Galápagos Islands.

An aggregation of green sea urchins at Pinta Island.

Toxopneustes roseus (A. Agassiz, 1863)   **Flower Sea Urchin**

Overall diameter 13 cm (5.1 in)
This distinctive urchin is covered with large pedicellariae which are held open, resembling numerous red-centered flowers. The spines are very short. The pedicellariae are venomous and should not be touched. The toxin produced by the pedicellariae paralyzes small animals (crabs and fish) but sea stars, some species of which feed on the flower urchin, are largely immune to the toxin. Like the green sea urchin, this species commonly displays a "covering reaction" by shielding itself with small rocks, bits of shells and other debris.
Habitat & range: Rocky substrate; low intertidal to 20 m (65 ft). It occurs throughout the central archipelago but is nowhere common. Gulf of California to Ecuador and the Galápagos Islands.

This flower urchin has covered itself with pebbles.

The same urchin "cleaned" by the photographer.

Tripneustes depressus A. Agassiz, 1863   **White Sea Urchin**

Overall diameter 16 cm (6.3 in)
This is the largest Galápagos urchin. It is globose, with short, almost white spines. This urchin is conspicuous, often seen singly or in pairs perched atop submerged boulders that are covered with algal tufts and crustose coralline algae, its principal foods.

Habitat & range:
Infralittoral to 73 m (240 ft), in areas of high water movement but not in areas of high turbulence. Common, although populations vary considerably in density from year to year. Southern California to Ecuador and Galápagos Islands.

Family Echinometridae

Caenocentrotus gibbosus (Agassiz & Desor, 1846)   **Brown Sea Urchin**

Overall diameter 9.5 cm (3.7 in)

This is a short-spined brown urchin similar to the purple sea urchin (*Echinometra vanbrunti*) in appearance and habit but differing in coloration. The test is brown, sometimes greenish, the spines a deep bronze-green, their tips more or less reddish.

Habitat & range: Mainly lowest intertidal but also reported to 9 m (30 ft). This species flourishes in the western archipelago where it often occurs with *Echinometra vanbrunti* in hollow depressions in intertidal tuffstone walls. Panama to northern Chile and the Galápagos Islands.

Echinometra vanbrunti A. Agassiz, 1863   **Purple Sea Urchin**

Overall diameter 11 cm (4.3 in)

This species has a somewhat oval, stout test with sturdy, purple, violet or brown spines of intermediate length. It tends to seek out crevices or to bore cavities in lava walls where it safely weathers full impact of the surf, but it may occur also on exposed surfaces on calmer shores. It feeds on drifting and attached algae and microscopic boring algae. It is considered an indicator species for the lowest littoral zone. The oval disc may attain 70 mm (2.8 in) but usually is smaller. In the western archipelago this species is in places replaced by *Caenocentrotus gibbosus*, which closely resembles the purple urchin.

Habitat & range: Shore to 53 m (175 ft), on both high-energy and semi-protected beaches. Common in the central archipelago, less common in the western archipelago. Monterey, California to Ecuador and Galápagos Islands.

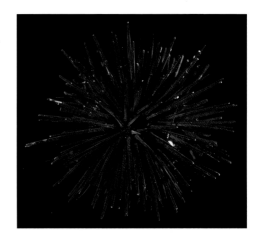

Echinometra oblonga (Blainville, 1825)   **Blunt-spined Echinometra**

Overall diameter to 80 mm (3.2 in)

This urchin closely resembles the purple sea urchin (*Echinometra vanbrunti*) in both appearance and habitat but is easily distinguished from the latter by the blunt-tipped spines and somewhat more elongate test. The color is a dark red or a very deep purple, the purple tinge more evident on the oral side.

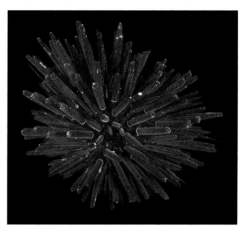

Habitat & range: Shallow water along rocky shores. This species does not bore depressions in rock (as does *E. vanbrunti*) but may inhabit depressions excavated by other urchins. Southern California to Ecuador and Galápagos Islands; also Indo-Pacific.

Family Mellitidae

Encope galapagensis A.H. Clark, 1946   **Galápagos Sand Dollar**

Length 13 cm (5.1 in)

This large Galápagos sand dollar is nearly circular in outline, with the edge perforated by 5 elongate holes, called lunules. A sixth lunule is longer and lies between two aboral petal-shaped areas. The spines of the sand dollar are used for locomotion, but are short and close-set, giving the animal a velvety appearance. This is the only sand dollar at all common in shallow water, although several other species have been collected along Galápagos shores by several expeditions. Sand dollars are deposit feeders that feed on sediment extracted from sand grains as they plow through the substrate.

Habitat & range: Flat sand bottom, low intertidal to 134 m (440 ft). Endemic to Galápagos Islands.

Family Schizasteridae

Agassizia scrobiculata Valenciennes, 1846   **Grooved Heart Urchin**

Length 2.5 cm (1 in)

This is a small, pale brown irregular urchin, the test about as high as it is wide. The five ambulacral areas radiate outward from a central pole. Four of the ambulacra form a conspicuous petal-shaped pattern (called "petaloid"); the two

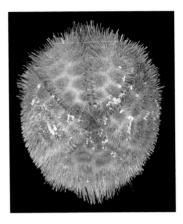

anterior petals are narrow and more than twice as long as the posterior pair. The single anterior ambulacrum is not part of the petaloid but forms a narrow band bearing tube feet on each side. The spines are short and rather brittle.

Habitat & range: Lowest intertidal to dredging depths, characteristically found burrowing just below the surface in sand; also occurs under rocks. Probably the most common heart urchin in Galápagos but seldom seen because of its burrowing habit. Gulf of California to Ecuador and the Galápagos Islands.

Brissus obesus Verrill, 1867   **Broad-keeled Heart Urchin**

Length 7 cm (2.8 in), width 5 cm (2 in)

This heart urchin is larger than *Agassizia scrobiculata* with an elongate test, the anterior end blunt and the posterior bluntly pointed. There are four ambulacral petals of nearly equal length on the aboral surface, the posterior pair only slightly longer than the anterior pair. Anterior to the anus on the oral (ventral) surface is a distinctive kidney-shaped area of darker coloration bordered by

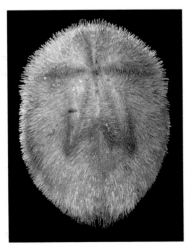

narrow bands of minute tubercles that support tiny ciliated spines. This is a burrowing species.

Habitat & range: Extreme low intertidal to dredging depths on sand and sand-mud bottoms and often under rock slabs. Probably uncommon, although it is difficult to judge population density because of its burrowing habit. From southern California to Panama and Galápagos Islands (new record).

Family Loveniidae

Lovenia cordiformis A. Agassiz, 1872   **Sea Porcupine**

Length 60 mm (2.4 in)

This heart urchin is easily identified by its long aboral, primary spines which curve back across the dorsal surface as though combed. The petaloid pattern on the dorsal surface is inconspicuous but the anterior ambulacrum lies in a well-marked V-shaped furrow, with the point of the V lying at the center of the petaloid pattern. The anus is positioned at the posterior end of the test and the mouth lies near the anterior end of the ventral surface. The ventral spines of the lower surface are enlarged with blunt tips, modified for burrowing forward through the sand.

Habitat & range: Extreme low littoral and shallow subtidal in sand and mud-sand bottoms where it lies buried, usually a few centimeters beneath the surface. Southern California, Gulf of California to Panama and the Cocos and Galápagos Islands.

Family Cassidulidae

Cassidulus pacifica (A. Agassiz, 1863)   **Pacific Cassidulus**

Length 60 mm (2.4 in)

This heart urchin has an oblong test with a highly convex, smoothly curved dorsal surface interrupted only by the anal furrow near the posterior end. The yellowish brown test is covered with an almost fur-like layer of close-set short spines. The petaloid pattern of the ambulacral areas is conspicuous. On the ventral side, extending from the extreme anterior to the posterior end, is a dagger-shaped area which crosses the mouth. This area is only sparsely covered with tiny spines, appearing as a polished surface. Like other heart urchins this is a burrowing species.

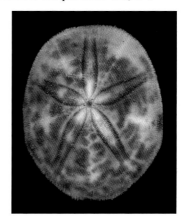

Habitat & range: Extreme low littoral and subtidal in sand and sand-mud bottoms. The species feeds by burrowing forward through sand, collecting small particles of organic matter on mucus strands which are passed to the mouth along the ambulacra. Gulf of California to Panama and the Galápagos Islands.

38

Family Clypeasteridae

Clypeaster rotundus (A. Agassiz, 1863) **Round Sea Biscuit**

Length to 17 cm (6.7 in)
Of the five species of *Clypeaster* recorded from Galápagos, only this species and *C. ochrus* are at all common. Both species of *Clypeaster* are large oval forms with thick, heavy tests covered with a dense fur of short spines. They resemble sand dollars of the genus *Encope* but are thicker, more arched in cross section, and lack edge perforations.

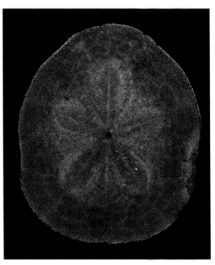

Clypeaster rotundus is the largest species of *Clypeaster* in Galápagos and easily distinguished from *C. ochrus* by its larger size and by its flatter test which is less arched on the dorsal surface and less concave on the ventral surface than *C. ochrus*.

Habitat & range: Found burrowing just beneath the surface of sandy bottoms. Common on sand bottoms, but seldom visible because of the burrowing habit. Gulf of California to Ecuador and the Galápagos Islands.

Clypeaster ochrus H.L. Clark, 1914 **Ocherous Sea Biscuit**

Length to 11 cm (4.3 in)
This sea biscuit is smaller than *C. rotundus*, seldom exceeding 9 cm (3.5 in) in length. The petaloid pattern on the dorsal surface and overall shape of the test is similar in both species but *C. ochrus* is much more arched on the dorsal side and more concave on the ventral side than *C. rotundus*.

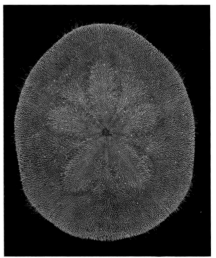

Habitat & range: On or in sandy bottoms where they burrow through the substrate to feed on particulate organisms. Gulf of California to Ecuador and the Galápagos Islands.

Other Sea Urchins
Reported from the Galápagos Islands

The following sea urchins, sand dollars, and heart urchins have been reported from the Galápagos intertidal and shallow-water habitats. They are considered uncommon or rare: *Arbacia incisa* (A. Agassiz, 1863); *Clypeaster elongatus* H.L. Clark, 1948; *Clypeaster europacificus* H.L. Clark, 1914; *Clypeaster speciosus* Verrill, 1870; *Encope micropora* L. Agassiz, 1841; *Mellitella stokesii* (L. Agassiz, 1841); *Brisaster latifrons* (A. Agassiz, 1898); *Brissiopsis pacifica* (A. Agassiz, 1898); *Meoma ventricosa grandis* Gray, 1851.

Sea Cucumbers

CLASS HOLOTHUROIDEA

*S*ea cucumbers are perhaps the oddest members of a phylum distinguished by strange animals. With their elongated body and warty surface, many sea cucumbers look remarkably like their vegetable namesake. Most Galápagos sea cucumbers belong to the genus Holothuria. *As a group, species of this genus are characterized by their large size, generally thick body walls, warty body surface, and microscopic ossicles, resembling tiny tables and buttons, embedded in the body wall (examples of these ossicles are illustrated in Appendix A). Many species have few predators and consequently often occur in exposed habitats. Some are plankton feeders. These extend bushy mucus-covered tentacles into the water to capture small organisms. Others, especially the larger species with simple tentacles, eat bottom sediment to extract the organic materials, egesting the waste. This method of feeding is typical of exposed forms such as* Holothuria kefersteini, H. portovallartensis, H. imitans, H. fuscocinerea *and* H. leucospilota. *There are no common names for the Galápagos sea cucumbers.*

Although sea cucumbers would appear to be easy meals for a host of predators, many have effective mechanical and chemical defenses that allow them to flourish undisturbed by predators such as fish. Several tropical sea cucumbers have long, threadlike tubules, called Cuvierian tubules, which can be ejected through the anus with considerable force when the cucumber is handled or disturbed by a potential predator. Fish or crabs become entangled in the tough, sticky white threads and immobilized. While only a few species of sea cucumbers produce sticky threads for defense, nearly all are protected by a potent toxin deposited in the sea cucumber's body wall. It is a highly effective defense against fish which learn to avoid sea cucumbers.

In Galápagos, only one species of sea cucumber, Stichopus fuscus, *is collected for human consumption. The presence of a large population of this species in the western archipelago has supported a thriving illegal fishery there. Sea cucumbers are prized as food in the Orient, where they are marketed as trepang or bêche-de-mer. The cucumbers are cooked and the protein-rich body walls dried and used in soups and other dishes. Because sea cucumbers are believed to be important processors of the bottom sediment, the illegal removal of large numbers of sea cucumbers is certain to have a negative impact on the productivity and diversity of the Galápagos benthic community. Already, scientists at the Charles Darwin Research Station have noticed increased accumulation of bottom sediment in areas where sea cucumber fishing has been intense.*

Most sea cucumbers can be identified from external appearance as detailed in the species descriptions, but there is often wide variation in coloration within a given species. Whether or not the animal shoots out sticky threads when handled is an additional diagnostic character. For some species, however, positive identification rests mainly upon examination of the microscopic ossicles in the body wall. For the benefit of scientists, extraction procedures and detailed descriptions of the body wall ossicles are provided in Appendix A.

Sea Cucumber Body Plan

Sea cucumbers have elongate, cylindrical, bilaterally symmetrical bodies with a mouth at one end encircled by 10 to 30 retractile feeding tentacles. The five ambulacral areas, so prominent in sea stars and sea urchins, are obscured in most sea cucumbers. Sea cucumbers lie on one side, the sole, which embraces three ambulacral areas and bears all the locomotory tube feet. The tube feet in many species lack suckers and are called papilliform; these have a nipple-like shape. The upper, or dorsal, surface (also called the dorsum) embraces the remaining two ambulacra. Here the tube feet are characteristically modified into warts and papillae. The body wall is soft or leathery because the endoskeletal plates, typical of their more calcified sister classes, are reduced to a profusion of microscopic ossicles. These minute ossicles, in curious geometric shapes, are important in the classification of sea cucumbers (see Appendix A).

The length measurements given in the species descriptions apply to relaxed specimens as they appear in their habitat.

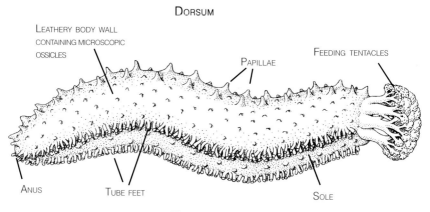

External anatomy of a sea cucumber.

Family Holothuridae

Holothuria difficilis Semper, 1868

Length 10-12 cm (4-4.7 in)

The body of this sea cucumber is stout, soft-walled, and covered with conspicuous pointed papillae irregularly spaced on the dorsal surface (these collapse when the animal is removed from the water). The ventral surface of larger specimens is flattened into a well-developed sole. This species is active at night and usually concealed under rocks during the day. It ejects sticky threads when handled. Body color is rich brick red with irregular black pigment blotches on the dorsum; some specimens are almost black. The ossicles (see Appendix A) consist of stout tables and large buttons. Because of its nocturnal habit, distinctive appearance, and ability to extrude sticky threads when handled,

this species can hardly be confused with any other Galápagos sea cucumber.

Habitat & range: Subtidal under rocks, exposed on rock surfaces at night, down to approximately 100 m (330 ft). Rarely occurs intertidally and in tide pools. Populations are of low density in the central archipelago but locally abundant in the western archipelago where it replaces *Stichopus horrens* as the dominant nocturnal species. From the eastern coast of Africa westward to Central America and Mexico and the Galápagos Islands.

H. difficilis foraging at night at Espinosa Point, Fernandina, 4 m (13 ft) depth

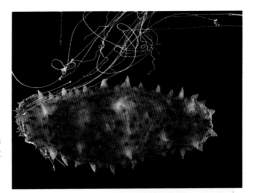

H. difficilis releasing sticky threads when disturbed, a characteristic that helps to identify this species.

Juvenile *H. difficilis*, 7 cm (2.8 in) long

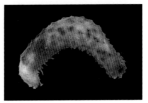

44

Holothuria arenicola (Semper, 1868)

Length 10-20 cm (4-8 in)

This common species is a spindle-shaped form of moderate size. Coloration is whitish-gray to light brown, sometimes distinctly yellow or golden, characteristically with two rows of distinct black spots on the dorsum. Sometimes distinct black spots are replaced by a scattering of small black spots. The ossicles (see Appendix A) consist of well-formed tables and buttons. This species lacks a distinct sole and does not release sticky threads when handled. It can be confused with *H. impatiens*, which sometimes bears a double row of black pigment spots on the dorsum, but *H. arenicola* is not nearly as warty as *H. impatiens*. The gold color of some specimens might lead to confusion with *H. portovallartensis*, but the latter is much softer-bodied, lacks a spindle shape, and lacks rows of spots on the dorsum.

Habitat & range: Almost always concealed beneath rocks. Found intertidally and subtidally. Common everywhere in Galápagos. Circumtropical.

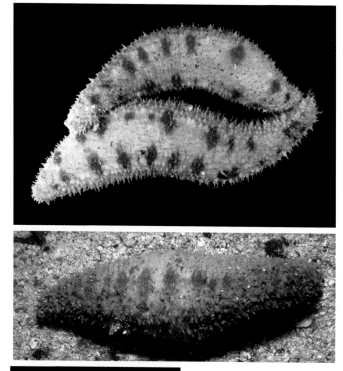

Two *H. arenicola* about 12 and 14 cm (4.7 and 5.5 inches) long. Note the double row of dark spots on the dorsum of each animal, a characteristic of the species.

H. arenicola exposed after removal of an overlying rock near Espinosa Point, Fernandina, 7 m (23 ft) depth.

With tentacles extended like miniature mops, *H. arenicola* explores the substratum for food.

Holothuria impatiens (Forskål, 1775)

Length 10-20 cm (4-8 inches)
This is a medium sized cucumber, elongate, often bottle shaped with a distinct "neck," especially when extended under water. The body wall is knobby due to pimple-shaped papillae, which are rough and rather firm to the touch. Coloration is usually brown with characteristic salt-and-pepper pigmentation between the papillae (wormlike pattern of black pigment on a white background when examined closely). The tube feet with white tips are positioned on tan-colored warts (papillae) which are much larger at the anterior end of the animal. This species ejects sticky threads when handled. The ossicles (see Appendix A) are tables and buttons.
Habitat & range: Intertidal and subtidal, almost always concealed beneath rocks; often covered with sand. Common everywhere in Galápagos. California to Ecuador and Galápagos Islands. Almost circumtropical in distribution.

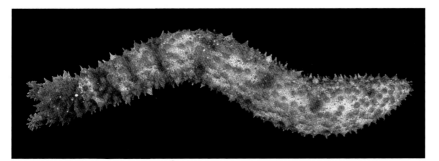

Holothuria impatiens, approximately 13 cm (5.1 in) long. Note the numerous knobby papillae of the body surface.

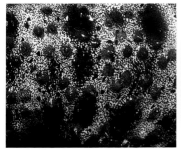

Closeup view of the body wall of *H. impatiens*, showing the typical "salt-and-pepper" pigmentation of the species.

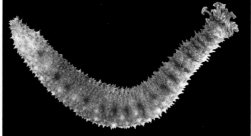

Juvenile *H. impatiens* approximately 10 cm (4 in) long.

Holothuria atra (Jaeger, 1833)

Length 20-30 cm (8-12 in)
This is a large, robust, jet-black sea cucumber with a smooth, cylindrical body. There are numerous soft tube feet on the ventrum and scattered, almost indistinguishable, papillae on the dorsum. This sea cucumber does not produce sticky threads. The ossicles (see Appendix A) consist of well-formed tables and small, scattered rosettes. This species is easily distinguished from *H. leucospilota*, another large, near-shore cucumber, by its smooth surface, thick body wall, and lack of sticky threads (*H. leucospilota* is covered with small papillae, has a thin, baglike body wall, and produces sticky threads when disturbed).

Habitat & range: Shallow water, lying exposed on coral and lava sand substrates. Uncommon, found only at Cartago Bay, Isabela, in our collections (first Galápagos record). Mozambique to Hawaii; Cocos, Clipperton, and Galápagos Islands.

ROBERT VAN SYOC

This large cucumber camouflages itself with sand while foraging (above). With the sand removed (right) its jet black color is revealed.

Holothuria hilla Lesson, 1830.

Length 15-20 cm (5.9-7.9 in)

This strikingly-colored medium sized cucumber resembles the closely-related *Holothuria impatiens*, but is easily distinguished from the latter by its brownish-yellow ground color and the large, pointed papillae arranged more or less in four series on the dorsal surface. The papillae are pale cream to white in color, giving the animal a distintive polka dot effect. The skin is thin and soft to the touch and almost translucent when the animal is extended. The ossicles (see Appendix A) are tables and buttons.

Habitat & range: Shallow water, concealed beneath rocks. Found throughout the central archipelago in Galapagos, but uncommon. From the east coast of Africa, Hawaii, and the Panamic region from the Gulf of California south to Ecuador and the Galápagos Islands.

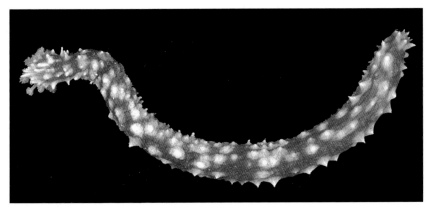

Adult *Holothuria hilla* from Darwin Bay, Tower Island.

Holothuria hilla from Tortuga Island. As with all sea cucumbers, coloration varies from site to site, and even among cucumbers at the same site.

Holothuria pardalis Selenka, 1867

Length 2-10 cm (0.8-4 in)
This is a small cucumber with papilliform tube feet in indistinct rows. Larger specimens have a distinct sole. The color is variable, light tan to nearly white with two rows of brownish-red pigmented areas on the dorsum. The tube feet on the dorsum appear as black spots on a tan field; the ventrum is pale. Juveniles have numerous small, rounded white warts scattered on a tan to nearly white ground color. Appendages are arranged in bands. The feet are much more papilliform anteriorly than on the rest of the body. Ossicles are stout tables with perforated bases and serrated edges, and buttons (see Appendix A). The extreme variability of coloration and body form makes this species difficult to identify from external appearance alone.

Habitat & range: Low intertidal to more than 30 m (100 ft). Specimens are typically exposed but partly covered with sand. Uncommon, although juveniles are relatively common under rocks in the low intertidal zone. Circumtropical except Atlantic.

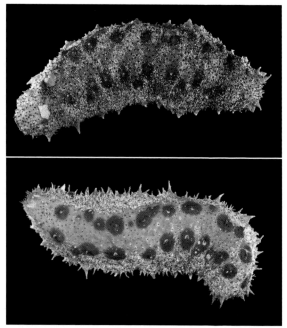

Two adult specimens of *H. pardalis*, 13 cm (5 in) long from the western archipelago (Elizabeth Bay and Banks Bay).

Small intertidal specimen of *H. pardalis*, approximately 7 cm (2.6 in) long.

Holothuria leucospilota (Brandt, 1835)

Length 20-40 cm (8-16 in)
This is a large, deep red or jet black sea cucumber with a baglike body covered with small papillae on the dorsum and small tube feet on the ventrum, these appearing as numerous small bumps when the animal is contracted and removed from the water. This species ejects copious sticky threads when handled or disturbed by a predator. The ossicles (see Appendix A) consist of irregular tables with incomplete discs, incomplete buttons, and occasional plates. It is present at certain sites along the east coast of Isabela, such as Cape Marshall, but evidently uncommon elsewhere. This may be the largest near-shore cucumber species in Galápagos, and is much larger than specimens of this species reported elsewhere in its range outside of Galápagos. Large specimens of this species and *Holothuria maccullochi* resemble each other, but the ossicles are completely different in the two species. The ossicles of this species are similar to those of *H. fuscocinerea*, but body form and color are altogether different.

Some specimens of *Holothuria leucospilota* are deep red in color.

Habitat & range: Shallow water to several meters where it lies exposed on coral and lava sand substrates. Eastern coast of Africa to Panamic region including Galápagos Islands.

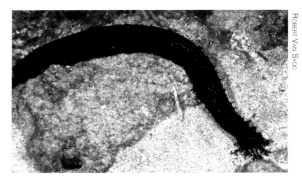

ROBERT VAN SYOC

A large adult *H. leucospilota*, approximately 30 cm (12 in) long, forages at 5 m (16 ft) depth near Cape Marshall, Isabela. The east coast of Isabela is a favored habitat of this species.

H. leucospilota releasing sticky threads in response to disturbance.

Holothuria fuscocinerea (Jaeger, 1833)

Length 20-32 cm (8-12.6 in)

This species is a large form, rather bag-like with a thin and flexible body wall. The general shape is long and cylindrical. It seldom contracts, even when handled. The color and surface features are distinctive: reddish brown to tan with conspicuous rounded pale papillae. Especially diagnostic are the elevated papillae which have a black spot or spots in their centers. The ventrum is pale. It remains exposed both day and night on rock surfaces, presumably protected from fish predation by the presence of toxins in the body wall. The body wall ossicles are tables and buttons (see Appendix A). This species is unlikely to be confused with any other Galápagos species.

Habitat & range: Low intertidal or shallow water to several fathoms; forages exposed on sand or rock bottoms. Found in association with *Holothuria kefersteini*. Southern Mexico to Galápagos, also Clarion and Hawaii Islands, and extends to east coast of Africa.

An adult *Holothuria fuscocinerea* from Rabida, depth about 8 m (26 ft). This species forages exposed both day and night.

Unlike many sea cucumbers, which contract when handled, *H. fuscocinerea* remains relaxed.

Holothuria imitans (Ludwig, 1875)

Length 12-20 cm (4.7-8 in)

Adults of this species are medium-sized, reaching a length of 20 cm (8 in), but most specimens are smaller. The body is typically spindle-shaped, rather thin-walled and soft. The base color is tan to rich brown with scattered small black spots on the dorsum, often with two rows of 10 to 12 large black blotches in each row. The papillae are white on the dorsum, prominent against the brown base color. The body wall ossicles are distinctive tables and rods, some plates, but no buttons (see Appendix A). This species is similar in external appearance to *H. arenicola* but *H. imitans* has white papillae and tube feet, whereas those of *H. arenicola* are the same color as the integument.

Habitat & range: Intertidal and shallow water, concealed beneath rocks. Relatively common throughout Galápagos. Panamic including Galápagos.

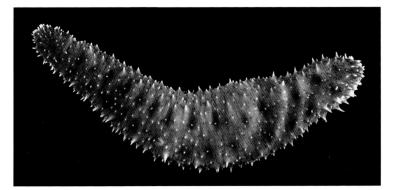

Holothuria imitans from Banks Bay, Isabela. Note the white tube feet and papillae, a diagnostic character.

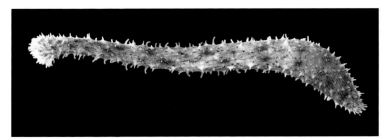

Small adult *Holothuria imitans*, approximately 16 cm (6.3 in) long. This specimen was living beneath a rock at Eden Island, depth 7 m (23 ft).

Holothuria kefersteini (Selenka, 1867)

Length 16-20 cm (6.3-8 in)

This is a large, robust species, commonly 20 cm (8 in) in length. The skin is thick, rough to the touch, and covered with randomly scattered small papillae and tube feet. Coloration is usually black with a deep red tinge in many specimens, but the color is variable. Juveniles are black and covered with pointed tubercles. This species remains exposed day and night, although well camouflaged with adhering sand in both intertidal and subtidal habitats. It does not release sticky threads when disturbed. The body wall ossicles are tables, rods, and perforated plates, but no buttons (see Appendix A). *H. kefersteini* resembles *H. leucospilota* in external appearance but *H. leucospilota* reaches a larger size and ejects sticky threads when disturbed. In the intertidal habitat it might be confused with *H. portovallartensis* or *H. theeli* but these two cryptic intertidal species lack the deep red background coloration of many (but not all) *H. kefersteini* specimens. Juveniles closely resemble juveniles of *H. theeli* and share the same intertidal habitat; only ossicle examination can safely distinguish the two species.

Habitat & range: Low intertidal and shallow water, typically exposed on coral sand bottoms and covered with sand that adheres to the body, providing some degree of camouflage. Often the dominant species on white sand bottoms of the central archipelago, sometimes in densities of several individuals per square meter. Mozambique to Hawaii; Cocos, Clipperton, and Galápagos Islands.

Holothuria kefersteini forages while exposed but partly camouflaged with sand and coral debris. It is one of the most common sea cucumbers in Galápagos.

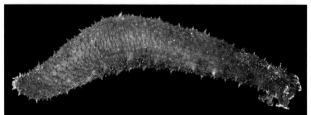

The deep red color of *H. kefersteini* is revealed when the animal is cleaned of adhering debris.

Holothuria theeli (Deichmann, 1938)

Length 8-13 cm (3.2-5.1 inches)
This medium-sized intertidal sea cucumber has a soft body wall which becomes firm in strongly contracted individuals. Well-developed tube feet and papillae are scattered across the surface, giving the animal a furry appearance. There are up to 20 bushy tentacles which are readily extended when a collected specimen is submerged. Juveniles tend to keep the tentacles extended even when handled. Adults are black with a hint of deep red in some specimens; the color of juveniles chocolate brown to nearly black. The body wall ossicles consist almost exclusively of small, irregular plates with a variable number of holes; no tables (see Appendix A). This species does not eject sticky threads. It is found in association with *H. portovallartensis*, which it closely resembles. In juveniles both external appearance and body wall ossicles are nearly identical in the two species.

Habitat & range: Intertidal at all levels, always concealed beneath rocks. It occasionally occurs in the high littoral zone, the only Galápagos sea cucumber to be found at this level. Preference, however, is for the mid- and low-littoral zones. Juveniles are usually rather firmly attached to the underside of rocks by tube feet. Outside of Galápagos this species has been reported to 60 m (200 ft). Gulf of California to Peru and Cocos and Galápagos Islands.

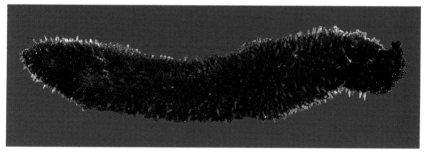

Adult *Holothuria theeli*, about 12 cm (4.7 in) long

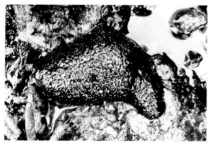

Juvenile *Holothuria theeli*, about 5 cm (2 in) long.

An overlying rock was removed to expose this *Holothuria theeli* in the low intertidal zone. In Galápagos this species is strictly intertidal in its distribution, always concealed beneath rocks.

Holothuria portovallartensis Caso, 1954

Length 8-14 cm (3.2-5.5 in)
This intertidal sea cucumber resembles *H. theeli* and is easily confused with it. Adults are of moderate size, soft-skinned, elongate and rather uniform in diameter except when contracted. There are numerous tube feet on the ventrum and small, soft papillae on the dorsum; relaxed individuals in water appear furry. The 12-20 tentacles are large, often bushy. The coloration of adults is variable, ranging from yellow (beneath the papillae) to varying shades of dark grey, dark brown to almost black. The ventrum is often yellow-green. Young specimens are chocolate brown. The body wall ossicles (see Appendix A) consist of plates and broad bars, often curved, with numerous marginal or terminal holes. Juveniles of this species and *H. theeli* are nearly identical in both external appearance and in the form of the body wall ossicles. Adults are easily distinguished from *H. theeli* by the yellow color of the ventrum and by the lacelike appearance of the body wall ossicles.

Habitat & range: Intertidal, always concealed beneath beach stones. Outside of Galápagos this species has been reported to 120 m (395 ft). Mexico to Peru and Galápagos Islands.

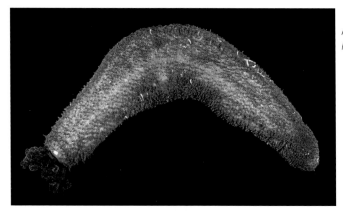

Adult *Holothuria portovallartensis.*

Two specimens of *H. portovallartensis*, exposed by the removal of an overlying rock, show variation in coloration. This species lives in the low intertidal zone.

Holothuria maccullochi Deichmann, 1958

Length 18-26 cm (7.1-10.2 in)
This uncommon, large sea cucumber, previously unreported from Galápagos, has an elongate, soft body of mostly uniform diameter when relaxed. The ground color of the skin is a deep, brick red. The tube feet on the ventrum are yellow and arranged in irregular double rows. On the dorsum the tube feet are scattered, small, and papilliform. The body wall ossicles are large and small tables (see Appendix A). While foraging, this species typically is covered with sand, similar to *H. kefersteini*.

Habitat & range: Intertidal and shallow water. This species is uncommon but occurs throughout the central archipelago. Outside of Galápagos it is reported as a "fugitive" species, usually found concealed under rocks, sand, or coral fragments. Gulf of California to Colombia and Galápagos Islands.

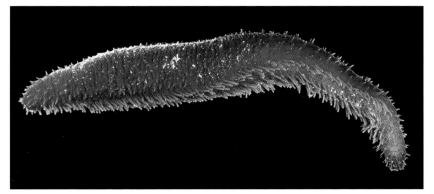

Holothuria maccullochi. This species is distinguished by its yellow tube feet and deep red ground color of its skin.

Holothuria rigida (Selenka, 1867)

Length 7-10 cm (2.8-4 in)
This medium-sized sea cucumber has a somewhat flattened body and rigid, tough skin covered with small papillae. Color white to light yellow. The ossicles (see Appendix A) consist of knobbed buttons and large tables with numerous teeth connected together to form a large, multilayered, hemispherical mass. The buttons resemble, and are easily confused with, the knobbed buttons of *Neothyone gibber.*

Habitat & range: Subtidal among rocks and on rock/ sand bottom. Uncommon. East coast of Africa and the Red Sea to the Panamic region and Galápagos Islands.

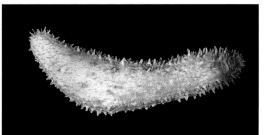

56

Family Stichopodidae
Stichopus fuscus (Ludwig, 1875)
Length to 25 cm (10 in)
This large sea cucumber is easily recognized by its thick, firm body wall bearing large, blunt papillae on the dorsum and three bands of cylindrical tube feet on the ventrum. The sole is distinct and well-formed. Coloration is light brown to chocolate brown with light tan papillae. This species produces ample slime when removed from water and handled, but does not eject sticky threads. The body wall ossicles consist of tables and C-shaped rods (see Appendix A). On average, this species reaches sexual maturity at about 160 g (19-20 cm [7.5-7.9 in] long). Peak spawning occurs in December through April although some spawning occurs throughout the year. The pepinos fishery in Galápagos is based exclusively on this species.

Habitat & range: Exposed on rock and sand substrates; shallow water to about 40 m (130 ft). Populations are composed of large individuals in rather high densities, especially in the cooler waters of the western archipelago. Tropical west coast of Mexico, Central, and South America; also Cocos, Socorro, and Galápagos Islands.

Adult *Stichopus fuscus*. Although found throughout Galápagos, this species flourishes especially in the colder waters of the western archipelago.

Juvenile *Stichopus fuscus*.

Stichopus horrens Selenka, 1867

Length to 30 cm (11.8 in)

This nocturnal species, previously unreported from Galápagos, is one of the largest Galápagos sea cucumbers, reaching 30 cm or more in length when active at night. It contracts to half or less its active length when inactive and concealed during the day. The body bears large, pointed papillae, called tubercles, which are fully extended in foraging animals at night. The tubercles are contracted and knoblike on dormant animals during the daytime. There are three columns of tube feet on the distinct sole. Coloration is highly variable, ranging from shades of deep red, orange, tan, brown , olive-green, to nearly black. The dorsum is often splotched with a medley of two, three, or four different colors. The body wall ossicles consist of small and large tables, rosettes, and C-shaped bodies (see Appendix A). The toxicity of this species to potential predators (fish and gastropods) is probably low, and it is known to escape predator attack by shedding a piece of body wall and moving rapidly down current. Despite being a common species, it was never reported by earlier expeditions to Galápagos which did most collecting during the daylight hours by dredging.

Habitat & range: Common in rocky areas of the central archipelago, subtidal 5-20 meters (16-66 ft). This species is cryptic and dormant during the day, contracting and wedging itself under rocky ledges. It is active and exposed only at night. Central Pacific, including the Philippines, Carolines, Fiji, Samoa, Society Islands, Hawaii, and Galápagos Islands.

Night-foraging *Stichopus horrens.*

Three *Stichopus horrens* together under a rocky ledge where they remain contracted during the day.

Stichopus horrens is highly variable in coloration.

58

Family Sclerodactylidae
Neothyone gibber (Selenka, 1867)

Length 5-6 cm (2-2.4 in)
This species is a small, stiff, thick-skinned form, the body usually strongly contracted with oral and caudal ends curved upward. The ten tentacles remain extended when handled. Coloration of the mid-body is light gray to gray-green, with anal and oral ends rich brown; the tentacles are dark brown. The body wall ossicles are faintly knobbed buttons (see Appendix A), quite unlike buttons of any species of the genus *Holothuria*.

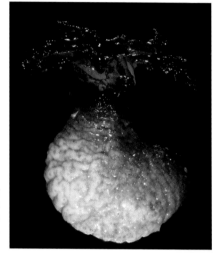

Habitat & range: Under rocks at subtidal depths, sometimes clinging to rocks but often floating free when overlying rocks are removed. This species has not previously been reported from Galápagos. Mexico to Peru and Galápagos Islands.

Neothyone gibber, a small and relatively uncommon sea cucumber in Galapagos, is a suspension feeder that ingests planktonic animals and plants and particles of detritus.

Family Phyllophoridae
Pentamera sp.

Length 10-15 cm (4-6 in)
This holothurian, tentatively placed in the genus *Pentamera*, was first collected at Caamaño Island (a small island southeast of Puerto Ayora on Santa Cruz) in 1995 by personnel of the Charles Darwin Research Station. Subsequently we have collected additional specimens of the species at other sites in the central and western archipelago. The animals are distinguished externally by their golden yellow color and long, active tube feet that are in constant motion. The ossicles (see Appendix A) consist entirely of tables with a round, perforated disc bearing a short, four-legged spire which is often incomplete. This holothurian is a new record for Galápagos and may be a new species.

Habitat & range: Subtidally under rocks. Possibly endemic to Galápagos.

Appendix A

Identification of Sea Cucumbers
by Examination of Body Wall Ossicles

Students of sea cucumber taxonomy base their classifications on several features of sea cucumber anatomy, of which by far the most important are the shapes of the microscopic skeletal ossicles that are embedded in the body wall of all sea cucumbers. The ossicles are broadly classified by shape. **Tables** are formed of a perforated disc on which rests a spire composed of usually four rods joined together by crossbars. **Buttons** are flattened, elongate, perforated bodies that may be smooth or knobbed. **Rosettes** are tiny ossicles formed of branching rods in a single plane. **Rods** are elongate bars, sometimes perforated on the ends, that serve as supporting structures in tube feet and tentacles. Sievelike perforated **plates**, often quite large in comparison to the ossicles just described, are common in many sea cucumbers, and usually do not serve a diagnostic purpose.

Extraction and Preparation Procedure

The extraction and preparation of the ossicles for examination is done easily with simple supplies: small test tubes, a medicine dropper, blank microscope slides and cover slips, and commercial bleach (e.g. Clorox). Examination of the ossicles requires a compound microscope.

• Slice a thin piece approximately 1 cm square and a millimeter thick from the body wall of the sea cucumber with a sharp knife or razor blade. It is not necessary to slice deeply into the body wall; the sea cucumber suffers no permanent harm from this procedure and may be returned to its environment.
• Drop the slice into a glass container (e.g., small test tube) and add a small amount (1-3 ml) of commercial bleach. Allow to stand for 30 minutes to an hour until the tissue is dissolved and the ossicles have drifted to the bottom of the test tube as a fine white sediment. (Do not expose bleach to direct sunlight, in which it rapidly decomposes.)
• Remove the supernatant carefully with a medicine dropper or pipette without disturbing the sediment and add clean water.
• When the sediment has settled after a few minutes, transfer some of the white sediment to a blank microscope slide, add a cover slip, and examine at X100 power with a compound microscope.
• If the body wall sample is to be preserved for later preparation, store in 70% alcohol, never formalin, because the latter will destroy the ossicles.

Identification

Holothuria (Microthele) difficilis Semper, 1868

Ossicles: Tables and buttons. Tables well-developed, stout, with smooth round disc bearing 8-10 holes rather evenly spaced, usually with smaller holes interspaced between the larger holes; table disc diameter average about 70 μm, height average about 55 μm; spire of moderate height, ending in a cluster of small teeth, these usually arranged in a square. Buttons from ventral integument large, oval, thin, flat, variable in size, with 3-7 pairs of small holes; button length average 115 μm, larger than buttons from any other common Galápagos species (important diagnostic character);
buttons mostly absent from dorsal integument. Buttons grade into small perforated plates of variable size and shape.

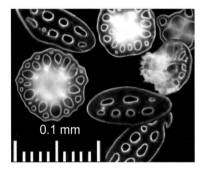

Similar species: The tables are similar to those of *H. arenicola* and *H. impatiens* but the buttons are much more irregular and larger in *H. difficilis* than either of those species.

Holothuria (Thymiosycia) arenicola (Semper, 1868)

Ossicles: Tables and buttons. Tables with low spire bearing a variable number of short teeth. Disc of table perforated by usually 8 holes but often more and sometimes less; holes of varying size. Table disc diameter 55-65 μm, spire height 46-50 μm. Buttons of moderate size and irregular, button length 45-50 μm; characteristically with 3 pairs of holes but often more; occasional plates look like large buttons bearing many holes. Short rods usually present with perforated edges.

Similar species: The ossicles are easily confused with *H. impatiens* but the tables are smaller with more rounded bases and have irregular rounded holes; buttons more irregular and smaller than *H. impatiens*; often necessary to measure table base diameter and button length to distinguish the two species by ossicles alone.

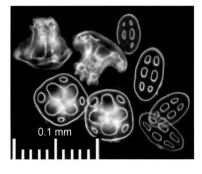

Holothuria (Thymiosycia) impatiens (Forskål, 1775)

Ossicles: Tables and buttons. Tables are short and stout with 8 large marginal holes in a squarish base, each marginal hole about as large as the central hole (the holes are more uniform in size than in *H. arenicola*); tables with a wreath of short spines on top. Table base diameter 70-85 µm, spire height 50-55 µm. Buttons are smooth and regular, usually with six large holes, larger (70-85 µm long) and more regular than in *H. arenicola*. Feet with end plate, smaller in the dorsal appendages. Supporting rods curved, with terminal holes, occasionally with lateral holes, giving them the form of large buttons.

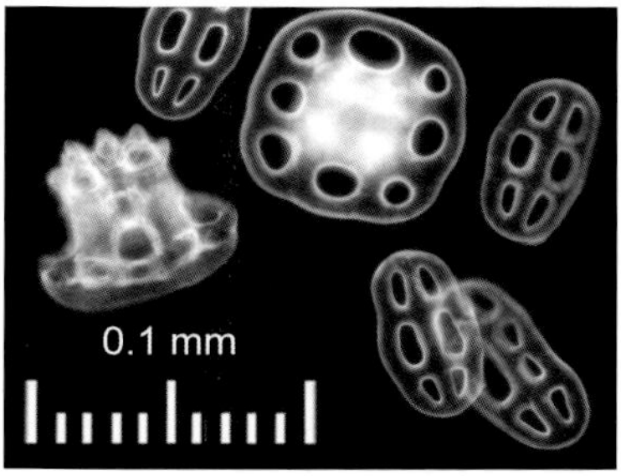

Similar species: Ossicles easily confused with those of *H. arenicola* but the tables are larger with squarish bases (those of *H. arenicola* are rounded) and have more regular perforations; buttons are more regular and larger than those of *H. arenicola*.

Holothuria (Ludwigothuria) atra (Jaeger, 1833)

Ossicles: Tables and rosettes. Tables with small disc and tall spire with four pillars ending in 12 well-formed teeth. The inner layer of skin consists of rosettes, usually bifurcate, but forms are variable.

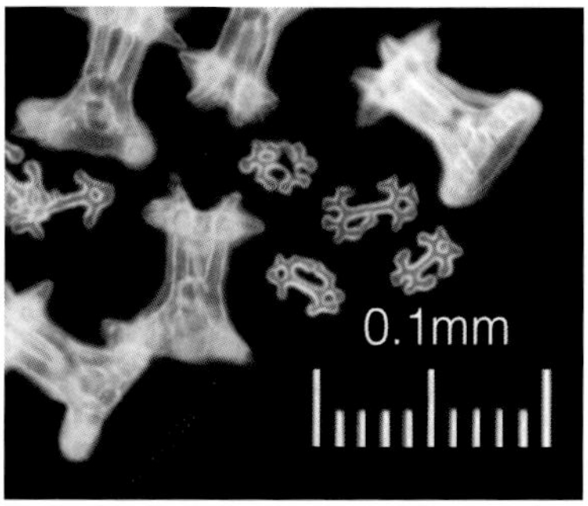

Similar species: The tables closely resemble those of *H. kefersteini*, but H. *kefersteini* lacks rosettes in the body wall.

Holothuria (Fossothuria) rigida (Selenka, 1867)

Ossicles: Large, hemispherical tables and knobbed buttons. Tables with a low spire bearing numerous small teeth that are connected together to form a large hemispherical mass. Buttons numerous and knobbed.

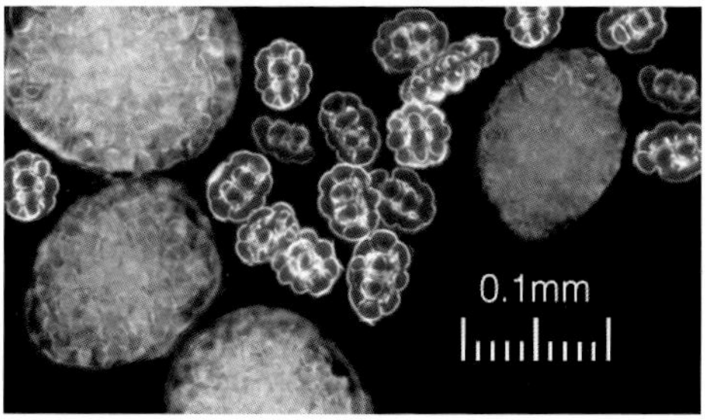

Similar species: The combination of hemispherical tables and knobbed buttons is unique. However, the buttons resemble those of the genus *Neothyone*.

62

Holothuria (Thymiosycia) hilla Lesson, 1830
Ossicles: Tables and buttons. Tables with a smooth-edged round disk with a complete circle of marginal holes, high tapering spire with four pillars and a few teeth on the top; top often incomplete. Table disc diameter 53-63 μm; table height 41-58 μm. Buttons elongate, smooth, with six elongate holes; button length 54-68 μm.
Similar species: Tables and buttons are very similar in shape and dimensions to those of *H. arenicola*, but the tables of *H. hilla* typically have a more smoothly-rounded disc and more perfectly round perforations in the disc than do the tables of *H. arenicola*. The two species are more easily distinguished by differences in external appearance than by differences in the ossicles.

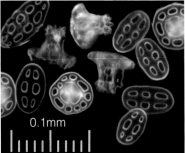

Holothuria (Lessonothuria) pardalis Selenka, 1867
Ossicles: Tables and buttons. Tables are stout with wide perforated base and serrated edge; spire low with 5-12 blunt teeth; incomplete tables common. Table base diameter 60-90 μm. Buttons vary from regular with 6-8 holes to (more commonly) irregular with 2-12 holes; frequently incomplete and twisted; button length 45-75 μm; buttons often appear in heaps. Curved rods in tube feet, with blunt ends or with one or more small holes on ends; some rods with expanded ends and perforated.
Similar species: None; combination of table and button form is diagnostic.

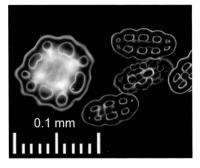

Holothuria (Mertensiothuria) leucospilota (Brandt, 1835)
Ossicles: Irregular tables and buttons and occasional plates. Tables with complete to incomplete discs, often reduced to four central holes and one to four marginal holes, but discs highly irregular. Tables may lack a spire, or have a low spire that may end in a flat crown of 8-12 blunt teeth. Buttons often incomplete, the typical button having two narrow slitlike holes at center and one or two pairs of minute holes at the ends. Scattered end plates, often quite large, perforated by numerous small round holes.
Similar species: Ossicles are easily confused with those of *H. fuscocinerea*.

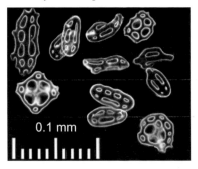

Holothuria (Mertensiothuria) fuscocinerea (Jaeger, 1833)
Ossicles: Tables and buttons. Tables are small and variable, disc often incomplete with typically 4 holes arranged in a square, sometimes with much smaller marginal holes; edge irregular but seldom spinous; spire low and usually absent. Disc diameter 27-36 µm. Buttons oblong, irregular; well-formed buttons have a pair of oblong central holes with a pair of small round holes at either end, but many buttons are incomplete. Buttons vary from 30-70 µm in length.
Similar species: The ossicles, especially the buttons, resemble closely those of *H. leucospilota*, but the tables are larger and more completely formed.

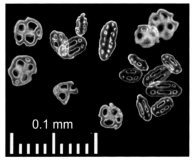

Holothuria (Semperothuria) imitans (Ludwig, 1875)
Ossicles: Tables and rods, some plates, no buttons. Tables tall, rather narrow with narrow base; spire topped by Maltese cross formed of 8 spines; sometimes small accessory spines between the 8 Maltese cross spines. Rods long and curved usually with perforated ends, sometimes with perforated lateral projections. Scattered perforated plates of variable size and form. Tables vastly outnumber other ossicles.
Similar species: None. The presence of tall, narrow tables topped by a Maltese cross of spines, and the absence of buttons, is a unique combination among Galápagos sea cucumbers.

Holothuria (Ludwigothuria) kefersteini (Selenka, 1867)
Ossicles: Tables, rods, and perforate plates, no buttons. Tables tall with small basal disc, spire with one cross-beam and four erect and 8 laterally-projecting teeth; tables often reduced. Rods relatively long, straight or curved, with perforated (sometimes lacelike) ends. Plates are small, rounded to elongate, perforated with small holes; edges serrated. Plates uncommon in surface skin samples.
Similar species: None.

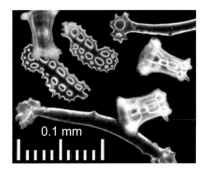

64

Holothuria (Selenkothuria) theeli (Deichmann, 1938)
Ossicles: No tables; spicules of plates with varying number of holes, edges of plates often with blunt teeth; also forked rods with holes in ends.
Similar species: The ossicles are similar to those of *H. portovallartensis*, especially among juveniles in which, with decreasing body size, they become increasingly difficult to distinguish.

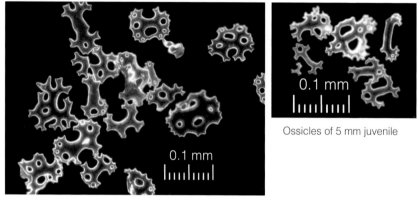

Ossicles of 5 mm juvenile

Ossicles of adult

Holothuria (Selenkothuria) portovallartensis Caso, 1954
Ossicles: Tables absent (except as vestiges in young individuals). Ossicles are plates and broad bars, often curved, with numerous marginal or terminal holes; surface of the margin often spinous. Ossicles of older individuals become increasingly lacelike.
Similar species: Ossicles of young individuals are somewhat more bar-like than those of *H. theeli*, but otherwise similar and easily confused.

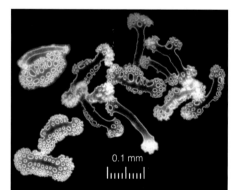

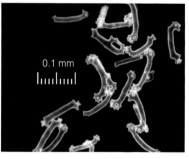

Ossicles of 9mm juvenile

Ossicles of adult

Holothuria (Irenothuria) maccullochi Deichmann, 1958
Ossicles: Large and small tables, and small, irregular, perforated discs. The disc
of large tables is 0.2 mm in diameter with numerous holes and bearing a tall
spire with cross beam near the base and four long, smooth pillars turning
outward near the tip, each tapering to a point. Small tables, few in number, with
perforated disc and short spire which is often reduced to 1-4 knobs or completely lacking. Dorsal papillae contain a few supporting rods with perforated
ends, and small perforated discs of various shapes.
Similar species: *Stichopus horrens* has tall ossicles, but the pillars unite at the
top into a point.

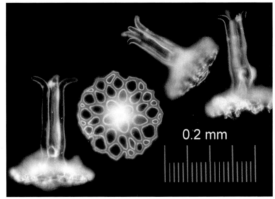

Stichopus fuscus (Ludwig, 1875)
Ossicles: Tables and C-shaped rods. Tables small, low and stout, about as wide
as tall, disc circular with 8-12 small holes, spire with wreath of small spines on
top. C-shaped rods (usually present) of varying lengths and numbers. Tube feet
contain endplates and perforated supporting plates.
Similar species: The tables are almost identical in shape to those of *S. horrens*,
but the latter species has
rosettes and large tack-like
ossicles, both lacking in *S.
fuscus.*

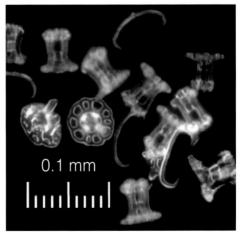

Stichopus horrens Selenka, 1867
Ossicles: Ossicles of large specimens consist of small and large tables, rosettes, and C-shaped bodies. Small tables are low and stout, disc square with characteristically 4 large and 4 smaller holes, 45-50 μm in diameter, spire 30-40 μm high and crowned with a wreath of blunt spines. Large tables are tack-like, 100-125 μm high, with disc diameter 110-125 μm and perforated with numerous holes. Rosettes are dichotomously branched rods. C-shaped bodies are 70-80 μm long.
Similar species: None. The combination of large and small tables, rosettes, and C-shaped bodies is unique among Galápagos sea cucumbers.

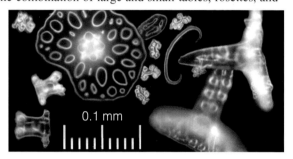

Neothyone gibber (Selenka, 1867)
Ossicles: Specially modified buttons that are faintly knobbed (unlike any of the genus *Holothuria*); some knobbed buttons bear an external cluster of spines and an inner "handle." Supporting rods with perforated ends also present.
Similar species: The ossicles resemble those of *Pentamera chierchia* but the two species are completely different in form and body color.

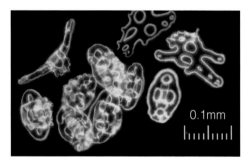

Pentamera sp.
Ossicles: Tables. Tables with small discs with dentate edges and variable number of holes. Spire short and incomplete.
Similar species: None.

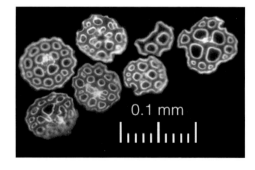

Appendix B

Record of Collection Sites for Photographs

Page numbers appear in bold type.

6 *Astropectin armatus*, Tagus Cove, 6 m; (lower)*Luidia foliolata*, Tagus Cove, 14 m, sand bottom.

7 *Luidia bellonae*, Caleta Iguana, Isabela, 10 m; (lower) *Nidorellia armata*, Pinzon, 8 m.

8 *Pentaceraster cumingi*, (left) Elizabeth Bay, Isabela, 4 m; (right) Cabo Marshall, Isabela, 5m; (lower) *Asterina* sp., Punta Espinosa, low littoral.

9 *Asteropsis carinifera*, Isla Pinta, 6 m; (lower) *Linckia columbiae*, (left) Elizabeth Bay, 12 m, (right) Pinzon, 14 m.

10 *Leiaster teres*, Cape Marshall, 8 m; (lower) *Pharia pyramidata*, Punta Pitt, San Cristobal.

11 *Phataria unifascialis*, Gardner Bay, Española, and Pinzon, 17 m; (lower) *Mithrodia bradleyi*, (left) Devil's Crown, Floreana, (right) Pinzon, 10 m.

12 *Acanthaster planci*, Darwin Island; (lower) *Heliaster cumingii*, Punta Mangle, Fernandina, intertidal.

13 *Heliaster solaris*, Punta Espinosa, Fernandina, collected June, 1974, by G. Wellington.

16 *Ophiophragmus* sp. Cape Douglas, 18 m; (lower) *Ophiactis simplex*, Academy Bay, low littoral.

17 *Ophiactis savignyi*, (left) Punta Pitt, San Cristobal, intertidal; (right) Academy Bay, mid littoral.

18 *Ophiothrix spiculata*, Punta Espinosa, Fernandina, intertidal; North Plaza, 2-10 m.

19 *Ophiothela mirabilis*, (left) Gardner Island at Floreana, 18 m; (right) Cape Douglas, 12 m.

20 *Ophiocoma aethiops*, (upper right) Kicker Rock, San Cristobal, 10 m; (lower) Gardner Bay, Española, 2 m.

21 *Ophiocoma alexandri*, Isla Darwin, north side, 12 m.

22 *Ophiocomella schmitti*, Punta Pitt, San Cristobal, intertidal.

23 *Ophionereis albomaculata*, (upper) Academy Bay, lower littoral; (lower) Roca Blanca (near Punta Albemarle), 6 m.

24 *Ophionereis perplexa*, Isla Wolf, 4 m; (lower) *Ophioderma teres*, Marchena, northwest side, 5 m.

25 *Ophioplocus hancocki*, both at Banks Bay, Isabela, 12 m.

29 *Astropyga pulvinata*, photo by Pablo Davila Jervis, Albany Island.

30 *Centrostephanus coronatus*, (upper) Gardner Bay, Española, 21 m; (lower) Isla Nameless, 20 m.

31 *Diadema mexicanum*, (upper) North Seymour, 2 m; (lower) Beagle Rocks, 8 m.

32 *Lytechinus semituberculatus*, (upper) Roca Blanca (near Punta Albemarle); (lower) Pinta, 3 m.

33 *Toxopneustes roseus*, Punta Cormorant, Floreana, 6 m; (lower) *Tripneustes depressus*, Tagus Cove, 8 m.

34 *Caenocentrotus gibbosus*, Punta Vicente Roca, 1 m, rock wall; *Echinometra vanbrunti*, Punta Vicente Roca, 1 m, rock wall.

35 *Echinometra oblongata*, near CDRS, low littoral; *Encope galapagensis*, Gardner Bay, Española, 2 m.

36 *Agassizia scrobiculata*, Gardner Island at Española, 17 m; *Brissus obesus*, Gardner Bay, Española, 2 m.

37 *Lovenia cordiformis*, Roca Blanca (near Punta Albemarle); *Cassidulus pacifica*, Roca Blanca (near Punta Albemarle), 8-10 m on sand.

38 *Clypeaster rotundus*, Tagus Cove, 14 m, sand bottom; *Clypeaster ochrus*, Banks Bay, 12 m, sand bottom.

43 *Holothuria difficilis*, (upper) Punta Espinosa, Fernandina, night dives, 5-7 m; (middle) Elizabeth Bay, 12 m; (lower) Piedras Blancas, Marchena, 4 m.

44 *Holothuria arenicola*, (upper) Punta Suarez, Española, 10 m, under rock; (middle) Punta Espinosa, Fernandina, 7 m, under rock; (lower) Genovesa, NW side, 5 m.

45 *Holothuria impatiens,* Gardner Island, Española, 18 m; (lower right) Kicker Rock, San Cristobal, 21 m.

46 *Holothuria atra*, Cartago Bay, 3 m.

47 *Holothuria hilla*, (upper) Darwin Bay, Genovesa, 9 m; (lower) Tortuga Island, 8 m.

48 *Holothuria pardalis*, (upper and middle) Elizabeth Bay, 12 m; and Banks Bay, 15 m; (lower) Tortuga Bay, Santa Cruz, intertidal.

49 *Holothuria leucospilota*, all from Cape Marshall, Isabela, 5-7 m.

50 *Holothuria fuscocinerea*, (upper) Rabida, 7 m; (lower) Isla Wolf, 5 m.

51 *Holothuria imitans*, (upper) Banks Bay, Isabela, 12 m; (lower) Eden, 7 m.

52 *Holothuria kefersteini*, (upper) Cape Marshall, Isabela, 5-7 m; (lower) Roca Este, San Cristobal, 21 m.

53 *Holothuria theeli*, all from Academy Bay near Charles Darwin Research Station, low littoral.

54 *Holothuria portovallartensis*, (upper) Academy Bay, low littoral; (lower) Tortuga Bay, Santa Cruz, low littoral.

55 *Holothuria maccullochi*, Punta Suarez, Española, 11 m; *Holothuria rigida*, Elizabeth Bay, 12 m.

56 *Stichopus fuscus*, (upper) Marchena; (lower) Punta Suarez, Española, 12 m.

57 *Stichopus horrens*, (upper) Punta Estrada, Santa Cruz, 10 m; (lower left), Rabida, 7 m; (lower right), Punta Vicente Roca, Isabela.

58 *Neothyone gibber*, James Bay, 6 m; (lower) *Pentamera* sp., Punta Suarez, Española, 12 m.

Appendix C

Guide to Further Reading

Following is a selection of references provided for those seeking additional information on the species included in this field guide. The Maluf number following each entry refers to numbered listings in L.Y. Maluf, 1988, in which geographic and depth ranges are given, as well as additional references.

Sea Stars

Astropecten armatus Gray, 1840. Maluf #2013
Boone (1928, as *Astropectin erinaceus*), Brusca (1980), Caso (1943, 1961), H. L. Clark (1910, 1913, 1940), Fisher (1906), Hopkins and Crozier (1966), Maluf (1988, 1991), Sladen (1889), Verrill (1867a, 1867b), Ziesenhenne (1937).

Luidia foliolata Grube, 1866. Maluf #2009
Caso (1961, 1979), Fisher (1911), Hopkins and Crozier (1966, as *Petalaster foliolata*).

Luidia bellonae Lütcken, 1864. Maluf #2002
Caso (1979a), H.L. Clark (1902), Steinbeck & Rickets (1941), Verrill (1867a, 1867b), Ziesenhenne (1937).

Nidorellia armata (Gray, 1840). Maluf #2073
Boone (1933), Brusca (1980), Caso (1943, 1961), H. L. Clark (1910, 1920a), Ely (1942), Maluf (1988, 1991), Sladen (1889), Verrill (1867a), Ziesenhenne (1937).

Pentaceraster cumingi (Gray, 1840) (Formerly *Oreaster occidentalis* Verrill, 1867). Maluf #2074
Brusca (1980), Caso (1943 as *Oreaster occidentalis*, 1961), H. L. Clark (1910), Fisher (1906), Maluf (1988, 1991), Sladen (1889), Verrill (1867a, 1867b), Ziesenhenne (1937).

Asterina sp.
Clark (1910), Madsen (1956), Sladen (1889).

Asteropsis carinifera (Lamarck, 1816). Maluf #2079
Caso (1961), Clark & Rowe (1971), Ely (1942), Kerstitch (1989), Ludwig (1905), Maluf (1988, 1991), Sladen (1889), Verrill (1867b).

Linckia columbiae Gray, 1840. Maluf #2082
Boone (1933), Brusca (1980), Caso (1948, 1961), H. L. Clark (1913, 1940), Hopkins and Crozier (1966), Maluf (1988, 1991), Sladen (1889), Verrill (1867b), Ziesenhenne (1937).

Leiaster teres (Verrill, 1871). Maluf #2081
Downey (1976), Jangoux (1980), Steinbeck & Ricketts (1941), Verrill (1867a, 1867b), Ziesenhenne (1937).

Pharia pyramidata (Gray, 1840). Maluf #2086
Boone (1933), Brusca (1980), Caso (1943), H. L. Clark (1910, 1913, 1920, 1940), Maluf (1988, 1991), Sladen (1889), Verrill (1867a), Ziesenhenne (1937).

Phataria unifascialis (Gray, 1840). Maluf #2088
Brusca (1980), Caso (1943, 1961), H. L. Clark (1910, 1913, 1940), Maluf (1988, 1991), Sladen (1889), Verrill (1867a, 1867b), Ziesenhenne (1937).

Mithrodia bradleyi Verrill, 1867. Maluf #2091
Brusca (1980), Caso (1944, 1961, 1963, 1979a), H. L. Clark (1910), Ely (1942), Maluf (1988, 1991), Pope & Rowe (1977), Sladen (1889), Verrill (1867a, 1867b).

Acanthaster planci (Linnaeus, 1758). Maluf #2093
Barham, Gowdy, & Wolfson (1973), Birkeland & Lucas (1990), Brusca (1980), Caso (1962), Dana & Wolfson (1970), Glynn (1974), Nishida & Lucas (1988), Porter (1972), Ziesenhenne (1935).

Heliaster cumingii (Gray, 1840). Maluf #2140
H. Clark (1907, 1920a), Madsen (1956), Maluf (1988, 1991), Sladen (1889), Verrill (1867a).

Heliaster solaris (A.H. Clark, 1920) (Formerly *H. multiradiata* (Gray, 1840). Maluf #2146
Boone (1933, as *H. multiradiatus*), A. Clark (1920), H. L. Clark (1902, 1907, 1920a), Maluf (1988, 1991), Sladen (1889), Verrill (1867b).

Brittle Stars

Ophiophragmus sp.
Ziesenhenne (1940), Nielsen (1932).

Ophiactis simplex (Le Conte, 1851). Maluf #3106
Brusca (1980), Maluf (1988, 1991), Morris, et al. (1980), Nielsen (1932), Steinbeck & Ricketts (1941), Verrill (1867a).

Ophiactis savignyi (Müller & Troschel, 1842). Maluf #3105
Brusca (1980), Caso (1951, 1961, 1979a, 1979b), Clark & Rowe (1971), H. Clark (1918, 1940), Ely (1942), Hendler, et al. (1995), Lyman (1882), Maluf (1988, 1991), May (1924), Verrill (1867a, 1867b).

Ophiothrix spiculata LeConte, 1851. Maluf #3115
Brusca (1980), Caso (1951, 1961, 1979a), H. L. Clark (1902, 1910, 1913, 1940), Lyman (1882), May (1924), Maluf (1988, 1991), Nielsen (1932), Verrill (1867a, 1867b), Ziesenhenne (1937).

Ophiothela mirabilis (Verrill, 1867). Maluf #3111
Verrill (1867a, 1867c), Nielsen (1932).

Ophiocoma aethiops Lütken, 1859. Maluf #3116
Boone (1933), Brusca (1980), Caso (1951, 1961, 1963, 1979a), H.L. Clark (1902, 1913, 1940), Kerstitch (1989), Lyman (1882), Maluf (1988, 1991), Nielsen (1932), Verrill (1867a, 1867b), Wellington (1975, p. 51), Ziesenhenne (1937).

Ophiocoma alexandri Lyman, 1860. Maluf #3117
Brusca (1980), Caso (1951, 1961, 1963, 1979a), H.L. Clark (1913, 1940), Kerstitch (1989), Lyman (1882), Maluf (1988, 1991), Nielsen (1932), Verrill (1867a, 1867b), Wellington (1975, p. 51), Ziesenhenne (1937).

Ophiocomella schmitti A.H. Clark, 1939. Maluf #3119
A. Clark (1939), Devaney (1970), Maluf (1988).

Ophionereis albomaculata E.A. Smith, 1877. Maluf #3125
A.Clark (1939), H.L. Clark (1902), Lyman (1882), Maluf (1988, 1991).

71

Ophionereis perplexa Ziesenhenne, 1940
Ziesenhenne (1940), A. Clark (1953).

Ophioderma teres (Lyman, 1860). Maluf #3136
Brusca (1980), Caso (1951, 1961), H.L. Clark (1902, 1940), Lyman (1882), Maluf (1988, 1991), Nielsen (1932), Verrill (1867a, 1867b), Wellington (1975, p. 51), Ziesenhenne (1955).

Ophioplocus hancocki Ziesenhenne, 1935. Maluf #3166
Ziesenhenne (1935), Steinbeck and Ricketts (1941).

Sea Urchins

Eucidaris thouarsii (Valenciennes, 1846). Maluf #4003
Boone (1933), Brusca (1980), Caso (1948, 1978), H. L. Clark (1948), Glynn (1979), Kerstitch (1989), Wellington (1975).

Astropyga pulvinata (Lamarck, 1816). Maluf #4015
Brusca (1980), Caso (1948, 1978), H. L. Clark (1948), Maluf (1988, 1991), Mortensen (1940).

Centrostephanus coronatus (Verrill, 1867). Maluf #4016
Brusca (1980), Caso (1978), H. L. Clark (1948), Kerstitch (1989), Maluf (1988, 1991), Wellington (1975, pp. 66, 224).

Diadema mexicanum A. Agassiz, 1863. Maluf #4017
Brusca (1980), Caso (1961, 1978), Maluf (1988, 1991), Mortensen (1940), Verrill (1867a), Wellington (1975, pp. 66, 114, 224).

Lytechinus semituberculatus (L. Agassiz & Desor, 1846). Maluf #4030
H.L. Clark (1948), Maluf (1988, 1991), Mortensen (1943), Wellington (1975, pp. 65, 114, 224, 257).

Toxopneustes roseus (A. Agassiz, 1863). Maluf #4031
Brusca (1980), Caso (1949, 1978), H.L. Clark (1948), Kerstitch (1989), Maluf (1988, 1991), Mortensen (1943), Wellington (1975, p. 224).

Tripneustes depressus A. Agassiz, 1863. Maluf #4032
Brusca (1980), Caso (1978), H.L. Clark (1948), Maluf (1988, 1991), Mortensen (1943), Wellington (1975, p. 224).

Caenocentrotus gibbosus (Agassiz & Desor, 1846). Maluf #4034
Boone (1933, as *Strongylocentrotus gibbosus*), H.L. Clark (1948), Maluf (1988, 1991), Wellington (1975, pp. 57, 99, 224, 257, 262, 332).

Echinometra vanbrunti A. Agassiz, 1863. Maluf #4036.
Brusca (1980), Caso (1948, 1978), H.L. Clark (1948), Kerstitch (1989), Maluf (1988, 1991), Mortensen (1943), Verrill (1867a), Wellington (1975, pp. 57, 58, 99, 224).

Echinometra oblonga (Blainville, 1825). Maluf #4035
Caso (1961, 1963, 1978), H.L. Clark (1912, 1948), Mortensen (1943), Ziesenhenne (1937).

Encope galapagensis A.H. Clark, 1946. Maluf #4051
A.H. Clark (1946), Maluf (1988, 1991), Wellington (1975, pp. 74, 126, 257).

Agassizia scrobiculata Valenciennes, 1846. Maluf #4074
Brusca (1980), H.L. Clark (1910, 1948), Clark & Rowe (1971), Kerstitch (1989), Maluf (1988, 1991), Mortensen (1951), Steinbeck & Ricketts (1941), Wellington (1975, p. 225).

72

Brissus obesus Verrill, 1867. Maluf #4081 (Synonym: *Brissus latecarintus*)
Brusca (1980), Caso (1983), H.L. Clark (1948), Hyman (1955), Maluf (1988, see *Brissus obesus*),
Mortensen (1951).

Lovenia cordiformis A. Agassiz, 1872. Maluf #4092
Brusca (1980), Caso (1961,1983), Clark & Rowe (1971), H.L. Clark (1910, 1913, 1940, 1948),
Mortensen (1951), Steinbeck & Ricketts (1941), Verrill (1971b), Ziesenhenne (1937).

Cassidulus pacifica (A. Agassiz, 1863). Maluf #4064
Caso (1949, 1961, 1983), H.L. Clark (1902, 1914, 1948), Kier (1962), Mortensen (1903), Wellington
(1975, p. 225).

Clypeaster rotundus (A. Agassiz, 1863). Maluf #4043
Agassiz (1872-73), Brusca (1980), Clark (1902, 1914, 1948), Mortensen (1948), Steinbeck and
Ricketts (1941), Verrill (1967a, as *Stoloniclypeus rotundus*).

Clypeaster ochrus H.L. Clark, 1914   Maluf #4042
Caso (1980), H.L. Clark (1914, 1940, 1948), Mortensen (1948).

Sea Cucumbers
Holothuria difficilis Semper, 1868. Maluf #5077
Caso (1963, 1966), Cherbonnier (1951), Clark & Rowe (1971), H. L. Clark (1920), Deichmann
(1958), Fisher (1907), Maluf (1991), Rowe (1969).

Holothuria arenicola (Semper, 1868). Maluf # 5085
Caso (1961), Cherbonnier (1951), H. L. Clark (1933), Deichmann (1930, 1958), Fisher (1907),
Hendler, et al. (1995), Maluf (1988, 1991).

Holothuria impatiens (Forskål, 1775). Maluf #5087
Boone (1933), Brusca (1980), Caso (1957), Cherbonnier (1951), Clark & Rowe (1971), H.L. Clark
(1913), Deichmann (1930, 1958), Hendler et al. (1995), Maluf (1988, 1991), Rowe (1969), Théel
(1886).

Holothuria atra (Jaeger, 1833)
Caso (1965b), Chao, Chen, and Alexander (1993), Clark and Rowe (1971), Conand (1993).

Holothuria hilla Lesson, 1830
Lesson (1830), Caso (1965a, as *Microthele zihautanensis*), Cherbonnier (1951, as *Holothuria
monacaria*), Clark & Rowe (1967, as *Holothuria monacaria*), H.L. Clark (1920, as *Holothuria
monacaria*), Deichmann, 1938, 1958, as *Holothuria gyrifer*), Fisher (1907, as *Holothuria
monacaria*), Pawson (1969), Rowe (1969).

Holothuria pardalis Selenka, 1867. Maluf #5074
Caso (1961), Cherbonnier (1951), Clark & Rowe (1971), Deichmann (1938, 1958), Fisher (1907),
Ludwig (1894), Maluf (1988, 1991), Rowe (1969).

Holothuria leucospilota (Brandt, 1835). Maluf #5076
Caso (1963, 1966), Cherbonnier (1951, as *Holothuria fusco-rubra*), Clark & Rowe (1971),
Deichmann (1958), Maluf (1988, 1991).

Holothuria fuscocinerea (Jaeger, 1833). Maluf #5075
Caso (1961), Cherbonnier (1951, as *Holothuria pseudo-zacae*), Clark & Rowe (1971), Deichmann
(1958), Théel (1886).

Holothuria imitans (Ludwig, 1875). Maluf #5082
Caso (1961, 1963, 1966), Cherbonnier (1951), Deichmann (1930, as *H. languens*; 1958), Maluf (1988, 1991).

Holothuria kefersteini (Selenka, 1867). Maluf # 5072
Boone (1933, as *H. kefersteinii*), Caso (1957, 1961, 1963, 1979a), H.L. Clark (1922, as *Stichopus kefersteinii*), Deichmann (1938, as *H. inornata*; 1958), Maluf (1988, 1991), Théel (1886).

Holothuria theeli (Deichmann, 1938). Maluf #5080
Caso (1961), Deichmann (1938, as *Holothuria marenzelleri* Ludwig var. *theeli* var. nov.; 1958), Ludwig (1894), Maluf (1988, 1991), Rowe (1969).

Holothuria portovallartensis Caso, 1954. Maluf #5079
Caso (1954, 1957, 1961), Deichmann (1938, as *Holothuria marenzelleri* Ludwig, var *theeli*, 1958), Maluf (1988, 1991).

Holothuria maccullochi Deichmann, 1958. Maluf #5073
Caso (1965), Deichmann (1958, as *Irenothuria maccullochi*), Rowe (1969).

Holothuria rigida (Selenka, 1867). Maluf #5070
Brusca (1980, as *Fossothuria rigida*), Clark & Rowe (1971, as *Cystipus rigida*), Deichmann (1930, 1958), Rowe (1969), Steinbeck and Ricketts (1941).

Stichopus fuscus (Ludwig, 1875). Maluf #5090
Brusca (1980), Caso (1957, 1961, 1966, 1967, 1979a), H.L. Clark (1910, 1922 as *Stichopus badionotus* Selenka), Deichmann (1938, 1958), Maluf (1988, 1991), Théel (1886).

Stichopus horrens Selenka, 1867. Maluf #5093
Clark & Rowe (1971), H. L. Clark (1922), Fisher (1907), Maluf (1988), Théel (1886).

Neothyone gibber (Selenka, 1867). Maluf #5040
Clark & Rowe (1971), H.L. Clark (1910), Deichmann (1938, 1941), Maluf (1988).

74

Selected References

Allen, R.K. 1990. Common intertidal invertebrates of southern California, rev. ed. Needham Heights, MA, Ginn Press.

Bakus, G.J. 1968. Defensive mechanisms and ecology of some tropical holothurians. Marine Biology, **2**:23-32.

Bakus, G.J. 1981. Chemical defense mechanisms on the Great Barrier Reef, Australia. Science, **211**:497-499.

Bakus, G.J. and G. Green. 1974. Toxicity in sponges and holothurians: a geographic pattern. Science, **185**:951-953.

Barham, E.G, R.W. Gowdy, and F.H. Wolfson. 1973. Acanthaster (Echinodermata, Asteroidea) in the Gulf of California. Fishery Bulletin of the National Oceanic and Atmospheric Administration **71**(4):927-942.

Boone, L. 1928. Echinoderms from the Gulf of California and the Perlas Islands. Bulletin of the Bingham Oceanographic Collection, Yale University, **2**(6):1-14.

Boone, L. 1933. Scientific results of cruises yachts "Eagle" and "Ara", 1921-1928. Coelenterata, Echinodermata, and Mollusca. Bulletin of the Vanderbilt Marine Museum, **4**:1-217, 133 pls.

Brusca, R.C. 1980. Common Intertidal Invertebrates of the Gulf of California, 2nd ed. Tucson, University of Arizona Press, 513 pp.

Caso, M.E. 1943. Contribución al conocimiento de los Astéridos de México. Tesis Profesional. Facultad de Ciencias, Universidad de México, 136 pp., 50 pls.

Caso, M.E. 1944. Estudios sobre Astéridos de México. Algunas especies interesantes de Astéridos litorales. Anales del Instituto de Biología, Universidad de México, Serie Ciencias del Mar y Limnología, **15**(1):236-259, pls. 1-7.

Caso, M.E. 1948. Contribución al conocimiento de los Equinoideos de México. II. Algunas especies de Equinoideos litorales. Anales del Instituto de Biología, Universidad de México, Serie Ciencias del Mar y Limnología, **19**(1):183-231.

Caso, M.E. 1949. Contribución al conocimiento de los Equinodermos de los litorales de México. Anales del Instituto de Biología, universidad de México, Serie Ciencias del Mar y Limnología, **20**(1-2):341-355, 6 pls.

Caso, M.E. 1951. Contribución al conocimiento de los Ofiuroideos de México, Serie Ciencias del Mar y Limnología, **22**(1):219-312.

Caso, M.E. 1954. Contribución al conocimiento de los Holothuroideos de México. Algunas especies de Holoturoideos litorales y descripción de una nueva especie, *Holothuria portovallartensis*. Anales del Instituto de Biología, Universidad de México, Serie Ciencias del Mar y Limología, **25**(1-2):417-442, 11 pls.

Caso, M.E. 1957. Contribución al conocimiento de los Holoturoideos de

México. III. Algunas especies de Holoturoideos litorales de la costa Pacífica Méxicana. Anales del Instituto de Biología, Universidad de México, Serie Ciencias del Mar y Limnología, **23**:309-338.

Caso, M.E. 1961. Estado actual de los conocimientos acerca de los Equinodermos de México. Tesis doctoral, Facultad de Ciencias. Universidad de México. 388 pp.

Caso, M.E. 1962. Estudios sobre Astéridos de México. Observaciones sobre especies Pacíficas del género *Acanthaster* y descripción de una subespecie nueva, *Acanthaster ellisii pseudoplanci*. Anales del Instituto de Biología, Universidad de México, Serie Ciencias del Mar y Limnología, **32**:313-331.

Caso, M.E. 1963. Estudios sobre Equinodermos de México. Contribución al conocimiento de los Equinodermos de las islas Revillagigedo. Anales del Instituto de Biología, Universidad de México, Serie Ciencias del Mar y Limología, **33**(1-2):367-380, 4 pls.

Caso, M.E. 1965a. Contribución al conocimiento de los Holoturoideos de México. Descripción de un nuevo subgenero del género Microthele y una nueva especie, Microthele (Paramicrothele) zihuatanensis. Anales del Instituto de Biología, Universidad de México, Serie Ciencias del Mar y Limnología, 35:105-114, + 3 plates.

Caso, M.E. 1965b. Estudios sobre Equinodermos de México. Contribucíon al conocimiento de los Holoturoideos de Zihautanejo y de la Isla ;de Ixtapa (primera parte). Anales del Instituto de Biología, Universidad de México, Serile Ciencias del Mar y Limnología. 36(1-2):253-291.

Caso, M.E. 1966. Estudios sobre Equinodermos de México. Contribución al conocimiento de los Holoturoideos de Zihuatanejo y de la Isla de Ixtapa (primera parte). Anales del Instituto de Biología, Universidad de México, Serie Ciencias del Mar y Limnologia, 36(1-2):253-291, 6 pls.

Caso, M.E. 1967. Contribución al estudio de los Holoturoideos de México. Morfología interna y ecología de *Stichopus fuscus* Ludwig. Anales del Instituto de Biología, Universidad de México, Serie Ciencias del Mar y Limnología, 37(1-2):175-182, 1 pl.

Caso, M.E. 1978. Los Equinoideos del Pacífico de México. Parte 1. Órdenes Cidaroidea y Aulodonta; Parte 2, Órdenlogía, México, Publicación Especial 1, 244 pp.

Caso, M.E. 1979a. Los Equinodermos de la Bahía de Mazatlán, Sinaloa. Anales del Centro de Ciencias del Mar y Limnología, México, **6**(1):197-368.

Caso, M.E. 1979b. Los Equinodermos (Asteroidea, Ophiuroidea y Echinoidea) de la Laguna de Términos, Campeche. Anales del Centro de Ciencias del Mar y Limnología, México, Publicación Especial 3, 186 pp.

Caso, M.E. 1980. Los Equinoideos del Pacífico de México. Parte 3. Órden Clypeasteroida. Anals del Centro de Ciencias del Mar y Limnología, México, Publicación Especial 4, 252 pp.

Caso, M.E. 1983. Los Equinoideos del Pacífico de México. Parte 4. Órdenes Cassiduloida y Spatangoida. Anales del Centro de Ciencias del Mar y

76

Limnología, México. Publicación Especial 6, 200 pp.

Chanley, J.D., R. Ledeen, J.Wax, R.F. Nigrelli, and H. Sobotka. 1959. Holothurin. I. Isolation, properties and sugar compounds of holothurin A. Journal of the American Chemical Society, 81:5180-5183.

Cherbonnier, G. 1951. Holothuries de l'Institut Royal des Sciences Naturelles de Belgique. Mémoires de l'Institut Royal des Sciences Naturelles de Belgique ser. 12, 41:1-64, 28 pls.

Clark, A.H. 1946. Echinoderms from the Pearl Islands, Bay of Panama, with a revision of the Pacific species of the genus Encope. Smithsonian Miscellaneous Collections, 106(5):1-ll, 4 pls.

Clark, A.M. 1920. A new name for Heliaster multiradiatus (Gray). Proceedings of the Biological Society of Washington, 33:183.

Clark, A.M. 1939. Echinoderms (other than holothurians) collected on the Presidential Cruise of 1938. Smithsonian Miscellaneous Collections, 98(11):1-18, 5 pls.

Clark, A.M. 1953. A revision of the genus Ophionereis (Echinodermata, Ophiuroidea). Proceedings of the Zoological Society of London, 123(1):65-94, 3 pls.

Clark, A.M. and F.W.E. Rowe. 1967. The identity of the species commonly known as Holothuria monacaria Lesson, 1830. Z.N.(S.) 1793. Bulletin of Zoological Nomenclature. 24:126-128.

Clark, A.M. and F.W.E. Rowe. 1971. Monograph of shallow-water Indo-West Pacific Echinoderms. London, Trustees of the British Museum (Natural History), 238 pp., 31 pls.

Clark, H.L. 1902. Papers from the Hopkins Stanford Galápagos expedition. 1898-1899. XII. Echinodermata. Proceedings of the Washington Academy of Sciences, 4:521-531.

Clark, H.L. 1907. The starfishes of the genus Heliaster. Bulletin of the Museum of Comparative Zoology at Harvard University, 51(2):25-76.

Clark, H.L. 1910. The echinoderms of Peru. Bulletin of the Museum of Comparative Zoology at Harvard University, 52(17):321-358, 14 pls.

Clark, H.L. 1912. Hawaiian and other Pacific Echini. Pedinidae, Phymosomatidae, Stomopneustidae, Echinidae, Temnopleuridae, Strongylocentrotidae and Echinometidae. Memoirs of the Museum of Comparative Zoology at Harvard College, 34(4):209-383, pls. 90-121.

Clark, H.L. 1913. Echinoderms from Lower California, with descriptions of new species. Bulletin of the American Museum of Natural History, 32:185-236, pls. 44-46.

Clark, H.L. 1914. Hawaiian and other Pacific Echini. The Clypeastridae, Arachnoididae, Laganidae, Fibulariidae and Scutellidae. Memoirs of the Museum of Comparative Zoology at Harvard College. 46(1):1-78, pls. 122-143.

Clark, H.L. 1917. Ophiuroidea. Report XVIII and XXX on the scientific results of the expedition of the "Albatross" to the tropical Pacific. Zoology at

Harvard University, **61**(12):429-453, 5 pls.

Clark, H.L. 1918. Brittlestars, new and old. Bulletin of the Musum of Comparative Zoology at Harvard University, **62**(6):26-338, 8 pls.

Clark, H.L. 1920a. Asterioidea. Report XXXII on the scientific results of the expedition of the "Albatross" to the tropical Pacific, 1904-1905. Memoirs of the Museum of Comparative Zoology at Harvard College, **39**(3):69-114, 6 pls.

Clark, H. L. 1920b. Holothuroidea. Report XXXIII on the scientific results of the expedition of the "Albatross" to the tropical Pacific, 1899-1900 and 1904-1905. Memoirs of the Museum of Comparative Zoology at Harvard College, **39**(4):121-154, 4 pls.

Clark, H.L. 1922. The Holothurians of the genus *Stichopus*. Bulletin of the Museum of Comparative Zoology at Harvard University, **65**:37-74, 2 pls.

Clark, H.L. 1933. Scientific survey of Puerto Rico and the Virgin Islands. Part 1. A handbook of the littoral echinoderms of Puerto Rico and the other West Indian Islands. New York Academy of Sciences, **16**:1-147, 7 pls.

Clark, H.L. 1940. Notes on echinoderms from the west coast of central America. Eastern Pacific Expeditions of the New York Zoological Society No. 21. Zoologica (New York), **25**:331-352, pls. 1-2.

Clark, H.L. 1948. A report on the echini of the warmer eastern Pacific, based on the collections of the Velero III. Allan Hancock Pacific Expeditions, **8**(5):225-352.

Conand, C., and M. Byrne. 1993. A review of recent developments in the world sea cucumber fisheries. Marine Fisheries Review, **55**(4):1-13.

Dana, T. and A. Wolfson. 1970. Eastern Pacific crown-of-thorns starfish populations in the lower Gulf of California. Transactions of the San Diego Society of Natural History, **16**(4):83-90.

Devaney, D. 1970. Studies on ophiocomid brittlestars. I. A new genus (*Clarkcoma*) of Ophiocominae with a reevaluation of the genus *Ophiocoma*. Smithsonian Contributions to Zoology, **51**:1-41.

Deichmann, Elisabeth. 1930. The holothurians of the western part of the Atlantic Ocean. Bulletin of the Museum of Comparative Zoology at Harvard University. **71**(3):43-226, pls. 1-24.

Deichmann, Elisabeth. 1938. Holothurians from the western coasts of Lower California and Central America, and from the Galápagos Islands: Eastern Pacific Expeditions of the New York Zoological Society. Zoologica (New York), **23**(18):361-387.

Deichmann, Elisabeth. 1941. The Holothuroidea collected by the "Velero" III during the years 1932 to 1938. Part 1. Dendrochirota. Allan Hancock Pacific Expeditions, **8**:61-153, pls. 10-30.

Deichmann, Elisabeth. 1958. The Holothuroidea collected by the "Velero" III and IV during the years 1932 to 1954. Part 2. Aspidochirota. Allan Hancock Pacific Expedition, **11**(2):249-331, pls. 1-9.

Ely, C.A. 1942. Shallow water Asteroidea and Ophiuroidea of Hawaii. Bulletin

78

of the Bernice Pauahi Bishop Museum, **176**:1-163.

Fisher, W.K. 1906. Starfishes of the Hawaiian Islands. Bulletin of the U.S. Fish Commission, 1903, Part 3, **23**:987-1130, 49 pls.

Fisher, W.K. 1907. The Holothurians of the Hawaiian Islands. Proceedings of the United States National Museum, **32**:637-744, pls. 66-82.

Fisher, W.K. 1931. Report on the South American seastars collected by Waldo L. Schmidt. Proceedings of the United States National Museum, **78**(16):1-10, 8 pls.

Glynn, P.W. 1974. The impact of *Acanthaster* on corals and coral reefs in the eastern Pacific. Environ. Conserv. **1**(4):295-304.

Glynn, P.W, G.M. Wellington, and C. Birkeland. 1979. Coral reef growth in the Galápagos: limitation by sea urchins. Science, **203**:47-49.

Hendler, G., J.E. Miller, D.L. Pawson, and P.M. Kier. 1995. Sea stars, sea urchins, and allies. Echinoderms of Florida and Caribbean. Washington, Smithsonian Institution Press.

Hopkins, T.S. and G.F. Crozier. 1966. Observations on the asteroid echinoderm fauna occurring in the shallow water of Southern California (intertidal to 60 meters). Bulletin of the Southern California Academy of Sciences. **65**(3):129-145.

Hyman, L.H. 1955. The invertebrates: Echinodermata. New York, McGraw-Hill Book Company, Inc.

Kerstitch, A. 1989. Sea of Cortex marine invertebrates. A guide for the Pacific coast, Mexico to Ecuador. Monterey, Sea Challengers, 114 pp.

Kier, P.M. 1962. Revision of the Cassiduloid Echinoids. Smithsonian Miscellaneous Collections, **144**(3):1-262, 44 pls.

Kropp, R.K. 1982. Responses of five holothurian species to attacks by a predatory gastropod *Tonna perdix*. Pacific Science, **36**(4):445-452.

Lesson, R.P. 1830. Centurie zoologique, ou choiz d'animaux rares, nouveaux ou imparfaitement connus. Paris, x, 244 pp., 80 pls.

Ludwig, H. 1894. Reports on an exploration off the west coasts of Mexico, central and South America, and off the Galápagos Islands, in charge of Alexander Agassiz, by the U.S. Fish Commission Steamer "Albatross," during 1891, Lieut. Commander Z.L. Tanner, U.S.N., commanding. XII. The Holothurioidea. Memoirs of the Museum of Comparative Zoology at Harvard University, **17**(3):1-183, 19 pls.

Ludwig, H. 1905. Asteroidea. Reports on an exploration off the west coasts of Mexico, Central and South America, and off the Galapagos Islands during 1891 by the "Albatross", A. Agassiz XXXV; Reports on the scientific results of the expedition to the tropical Pacific in charge of A. Agassiz aboard the "Albatross", from Aug. 1899 to Mar. 1900. VII. Memoirs of the Museum of Comparative Zoology at Harvard College, **32**:1-292, 35 pls.

Ludwig, H. 1920. Asteroidea. Report XXXII on the scientific results of the expedition of the "Albatross" to the tropical Pacific, 1904-1905. Memoirs of the Museum of Comparative Zoology at Harvard College, **39**(3):69-114, 6 pls.

Lyman, T. 1882. Report on the Ophiuroidea dredged by H.M.S. "Challenger" during the years 1873-76. Report of the Scientific Results of the Voyage of H.M.S. "Challenger" 1873-76, **5**(14):1-386, 48 pls.

Madsen, F. J. 1956. Reports of the Lund University Chile Expedition 1948-49. Asteroidea, with a survey of the Asteroidea of the Chilean Shelf. Acta Universitatis Lundensis (N.S.). **52**(2):1-53, 6 pls.

Maluf, L.Y. 1988. Composition and distribution of the Central Eastern Pacific echinoderms. Technical Report 2, Los Angeles, Natural History Museum of Los Angeles, pp. 1-242.

Maluf, L.Y. 1991. Echinoderm fauna of the Galápagos Islands. Chap. 16 in James, Matthew J., Galápagos Marine Invertebrates: Taxonomy, Biogeography, and Evolution in Darwin's Islands. New York, Plenum Press.

May, R.M. 1924. The ophiurans of Monterey Bay. Proceedings of the California Academy of Sciences, **13**(18):261-303.

Morris, R. H., D. P. Abbott, E. C. Haderlie. 1980. Intertidal invertebrates of California. Stanford, CA, Stanford University Press.

Mortensen, T. 1903. Echinoidea, Part 1. Danish Ingolf Expedition, **4**:1-193, 21 pls., 1 map.

Mortensen, T. 1940. A monograph of the Echinoidea. III. 1, Aulodonta. Copehagen: C. A. Reitzel, 370 pp., 77 pls.

Mortensen, T. 1943. A monograph of the Echinoidea. III. 2,3, Camarodonta. Copenhagen: C.A. Reitzel, 553 pp., 56 pls.; 446 pp., 66 pls.

Mortensen, T. 1948. A monograph of the Echinoidea. IV. 2, Clypeastroida, Clypeastridae, Arachnoididae, Fibularidae, Laganidae and Scutellidae. Copenhagen: C.A. Reitzel. 471 pp., 72 pls.

Mortensen, T. 1951. A monograph of the Echinoidea. V. 2, Spatangoida 2. Copenhagen: C.A. Reitzel, 593 pp, 64 pls.

Nielsen, E. 1932. Ophiurans from the Gulf of Panama, California, and the Strait of Georgia. Videnskabelige Meddelelser fra Dansk Naturhistorisk Forening, **91**:241-326.

Nishida, M. and J.S. Lucas. 1988. Genetic differences between geographic populations of the crown-of-thorns starfish throughout the Pacific region. Marine Biology, **98**:359-368.

Pawson, David L. 1969. Holothuroidea from Chile. Report No. 46 for the Lund University Chile Expedition 1948-1949. Sarsia **38**:121-146.

Pope, E.C. and F.W.E. Rowe. 1977. A new genus and two new species in the family Mithrodiidae (Echinodermata: Asteroidea) with comments on the status of species of *Mithroidia* Gray, 1840. Australian Zoologists, **19**(2):201-216.

Porter, J.W. 1972. Predation by *Acanthaster* and its effect on coral species diversity. American Naturalist, **106**(950):487-492.

Rowe, F.W.E. 1969. A review of the family Holothuriidae (Holothurioidea: Aspidochirotida). Bulletin of the British Museum (Natural History), Zoology, **18**(4):119-170.

Sladen, W. P. 1889. Asteroidea. Report of the scientific results of the voyage of the H.M.S. "Challenger" 1873-76, **30**:1-893, 117 pls.

Steinbeck, J., and E.F. Ricketts. 1941. Sea of Cortez. A leisurely journal of travel and research. New York: Viking Press, 598 pp.

Théel, H. 1886. Report on the Holothurioidea dredged by H.M.S. "Challenger" during the years 1873-1876, II. Report of the Scientific Results of the Voyage of H.M.S. "Challenger" 1873-76, **14**(39):1-290, 16 pls.

Verrill, A.E. 1867a. Notes on the echinoderms of Panama and west coast of America, with descriptions of new genera and species. Transactions of the Connecticut Academy of Arts and Sciences, **1**(2):251-322, pl. 10.

Verrill, A.E. 1867b. On the geographical distribution of the echinoderms of the west coast of America. Transactions of the Connecticut Academy of Arts and Sciences, **1**(2):323-338.

Verrill, A.E. 1867c. Notes on the Radiata in the Museum of Yale College, with descriptions of new genera and species. Transactions of the Connecticut Academy of Arts and Sciences, **1**(2):247-251.

Wellington, G.M. 1975. The Galápagos coastal marine environment. A resource report to the Department of National Parks and Wildlife, Quito. Unpublished manuscript, 357 pp.

Ziesenhenne, F. C. 1935. A new brittle star from the Galapagos Islands. Allan Hancock Pacific Expeditions, **2**(l):1-4, 1 pl.

Ziesenhenne, F.C. 1937. Echinoderms from the west coast of Lower California, the Gulf of California and Clarion Island. Zoologica (New York), **22**(15):209-239.

Ziesenhenne, F.C. 1940. New ophiurans of the Allan Hancock Pacific Expeditions. Allan Hancock Pacific Expeditions, **8**(2):9-42, 9 pls.

Ziesenhenne, F.C. 1955. A review of the genus *Ophioderma* M.&T. Essays in the natural sciences in honor of Captain Allan Hancock. Los Angeles: University of California Press, pp. 185-201.

Index to Scientific Names

S
savignyi, Ophiactis, 17
Schizasteridae, 36
schmitti, Ophiocomella, 22
Sclerodactylidae, 58
scrobiculata, Agassizia, 36
Selenkothuria, 64
semituberculatus, Lytechinus, 32
Semperothuria, 63
sertulifera, Astrometis, 13
simplex, Ophiactis, 16
solaris, Heliaster, 13
speciosus, Clypeaster, 39
spiculata, Ophiothrix, 18
Stichopodidae, 56
Stichopus, 56-57
stokesii, Mellitella, 39
superba, Luidia, 13

T
tabogensis, Ophiophragmus, 25
teres, Ophioderma, 24
teres, Leiaster, 10
theeli, Holothuria, 53
thouarsii, Eucidaris, 29
Thymiosycia, 60-62
Toxopneustes, 33
Toxopneustidae, 32
Tripneustes, 33

U
unifascialis, Phataria, 11

V
vanbrunti, Echinometra, 34
variegatum, Ophioderma, 25
ventricosa, Meoma, 39

A Field Guide to
Sea Stars
and other Echinoderms of
Galápagos

Cleveland P. Hickman, Jr.

A Field Guide to Sea Stars

and other Echinoderms of Galápagos

BY

Cleveland P. Hickman, Jr.

PHOTOGRAPHY BY

The author unless otherwise attributed

WITH ORIGINAL ART BY

William C. Ober, M.D. and Claire Garrison

Sugar Spring Press   Lexington, Virginia

1998

Library of Congress Catalog Number: 98-90489
ISBN: 0-9664932-0-6

Printed in the United States of America

email: hickman.c@wlu.edu

Sugar Spring Press
802 Sunset Drive
Lexington, Virginia, USA